Biodiversity & Environmental Management

Biodiversity & Environmental Management

B.D. Joshi
C.P.M. Tripathi
P.C. Joshi

APH PUBLISHING CORPORATION
4435-36/7, ANSARI ROAD, DARYA GANJ,
NEW DELHI-110 002

Published by
S.B. Nangia
A P H Publishing Corporation
7, Ansari Road, Daryaganj
New Delhi 110002
Ph.: 23274050
E-mail : aphbooks@gmail.com

2024

Printed at
Balaji Offset
Navin Shahdara, Delhi 110032

Preface

Man out of his earnest desire to make human life more comfortable and to meet out the growing requirements of the faster growing human population has converted this planet earth, revered as our mother earth, to a very fragile and nearly hazardous world. It is on account of three major dimensions: first man is over exploiting the natural resources of this earth, second he is synthesizing , using and leaving a lot of pollution causing, toxic waste material all over the earth, its atmosphere, lithosphere and hydrosphere and third he is very careless about myriad aspects of environmental sanctity and management. As a result we are fast loosing our precious bio-diversity, which has been the basis of human origin and evolution. Not only bio-diversity is being lost in its living physical form as a species, but that what ever is existing around us it is getting physically, genetically and physiologically distigured, on account of short term and long term impacts of our misuses and callousness towards environment and its basic components.

Therefore, the people who matter, all over the world, have become conscious regarding the earths ecological constituents and its bio-diversity, it is because the bio-diversity makes the earth habitable for human beings . We as academician and environment oriented researchers have greater responsibility to act in the direction of spreading awareness, educate people and future generations towards environmental ethics, values, importance, protection and impacts of careless misuse of earth resources, through education and creating fresh educational material in visual and print ftorm, with new and latest scientific information and researches, ft was with this background of our role that the *XIV National Seminar organized of Indian Academy of Environmental Sciences, Hardwar* was held at Department Of Zoology, Deen Dayal Upadhyay University, Gorakhpur , - during November 2007. Out of the presentations during the said seminar a total of 47 articles were selected for publication in book form. These articles have been divided in two volumes, on the basis their subjectivity. This volume entitled *Biodiversity & Environmental Management*, contains 22 articles, describing the latest researches m the related field and provides much required knowledge on myriad aspects of bio-diversity and environmental management. The second volume is entitled as *Environmental Pollution & Toxicology* containing 25 articles, is also being published simultaneously. The papers included here in are the original contributions of specialists in the respective disciplines, from all over the country. We express our thanks to the authors for their contributions. Special thanks are to *Dr. Kamal Kishore, and Mr. Vijaya Sharma*, my Ph.D. scholars in the Department of Zoology & Environmental Sciences, Hardwar, tor very painstakingly helping me in organizing these volumes. Thanks are also to the publisher, the APH publishing corporation for their prompt interests and co-operation at various levels.

B.D. Joshi
C.P.M. Tripathi
P.C. Joshi

Contents

1
Diversity in Community Structure of Coleopteron Insects in A Moist Deciduous Protected Forest in Uttarakhand, India

P. C. JOSHI AND NEELESH VASHISHTA
Department of Zoology and Environmental Science, Gurukula Kangari University, Hardwar-249404

Abstract

Species composition and community structure of order Coleoptera was studied for a period of two years in a moist deciduous forest of Uttarakhand State in India. Four study sites, which form part of a National Park were selected in manner that they vary in vegetational composition, altitude and anthropogenic encroachments. Sixteen species belonging to seven families of order Coleoptera were recorded during a period of two years. Family Scarabidae was the most dominant family in terms of number of species while, Coccinelidae in terms of number of total collected individuals. Greatest number of species were recorded from mixed forest with no disturbances i.e. site 1. Lowest number of individuals were recorded from the site which has plantation forest and was highly disturbed but it showed a high Shannon-Wiener diversity index (H) and also higher evenness as compared to other sites.

Introduction

Within a given habitat, physical and chemical characteristics often affect local insect distribution. Additionally the organization of insect species assemblage may be affected by the interaction of biotic and abiotic factors or by predation, parasitism and interspecific competition[1-2]. Coleopterans often constitute an important group of arthropods and are the most conspicuous insects relative to species composition, number, biomass and the extent of damage they cause to their host species. Species distribution depends mostly upon the presence or absence of the plant species and herbivorous insect species, which primarily feed on them. Outbreaks of forest insects also depend on external conditions like physically damaged stands and climatically altered soils[3-5]. Biotic factors and interspecific competition in particular may be specifically important in tropical habitats that harbor high species diversity. However, the population and community dynamics of tropical insect populations are poorly understood. Beeson[6] conducted a study on effect of vegetation and different abiotic factors such as temperature, humidity and rainfall on insect pests of forest ecosystems in India. Roonwal[7] and Mathur and Singh[8] compiled lists of insect pests of forest plants in India. However, little is known on species composition of Coleoptera from a moist deciduous forest, and collections have been made in the moist deciduous forests on relatively few occasions in the past, the relevant records of which are sparsely scattered through the literature and difficult to trace. However, a considerable amount of work on population dynamics of Coleoptera has been conducted in temperate and tropical communities of the world[9-15]. Studying and predictiong the spatial distribution of insects is one of the major concerns in forest entomology, and a major goal of ecology is to identify factors that predict species distribution and abundance, at both a geographical and a local scale. The present paper includes studies on species composition and community structure of Coleoptera with respect to vegetation communities and disturbances in a moist deciduous forest in India.

Materials and Methods

Study Area: The present study was conducted in a protected forest in the Motichur Forest Sanctuary of Rajaji National Park, Hardwar, India, located at 29°15' to 30°15' N and 77° 55' to 78° 30' E. Four sites each with an area of 3 ha. were selected within the protected forest, these fall in three categories representing habitats under different vegetation communities and levels of disturbances. All sampling sites were located within 12 km of each other and elevations of the study sites ranged between 360-440 m.

Site 1 (Mixed Forest With No Disturbance) : The site is a mix forest of *Shorea robusta* Gaertn.f. and *Mallotus philippinensis* muell-Arg. and falls in Motichur forest block at an elevation of 412 m. This site is free from all human disturbances. It has not been logged within the last twenty years and there was no grazing by cattle. A preliminary floral survey suggested that the understory vegetation in this site was diverse and dominated by herbaceous plants *Desmodium laxiflorum* D.C., *D. triangulare* (Retz) Merr. *Clerodendrum multiflorum* (Burm.f.) O.Kuntze. *Cynodon dactylon* (Linn) Pers.

Site 2 (Plantation Forest With Severe Disturbances) : This site is a plantation forest of *Tectona grandis* Linn.f., planted by State Forest Department approximately fifteen years earlier which falls in Koelpura forest block at an elevation of 370 m. This site is highly disturbed due to Gujjars, a muslim nomadic community, living in forests from last several hundered years. Gujjars collect fuel and fodder from this site, while their cattle roam in this site freely, destroying newly emerged plant saplings. The preliminary floral survey indicated a relatively depauperate understory community and was dominated by *Lantana* sp. and *Caressa opaca* Stapt.

Site 3 (Mixed Forest With Moderate Disturbance) : This site is again a mixed forest of *S. robusta and M. philippinensis* and is situated in Danda forest block at an elevation of 360 m. A small perennial water stream is also present in this site, due to which a part of this site looks marshy or water logged. This site is moderately disturbed by Gujjars as only three huts of Gujjars were present in this site and the number of cattle was also much less as compared to site 2. Thus this site represents a diverse forest ecosystem with a relatively modest level of current distrubance with a dense understory. The important constituents of understory vegetation include *Achyranthus aspera* Linn., *Ampelopteris prolifera* Copeland, *Arundinella* sp., *Calotropus. and Desmodium* sp.

Site 4 (Pure Sal Forest With No Ddisturbance) : This site is a pure dense forest of *S. robusta,* situated in Jamunkhata forest block at an elevation of 438 m. This site was selectively gogged some thirty to thirty five years earlier. At present it was totally free from all type of human disturbances. Based on preliminary floral survey, the understory plant community appeared to have a relatively low diversity but the density of shrubs is quite good in comparison to herbs.

Sampling of Entomofauna : Monthly sampling of insects was carried out, using sweep nets during 1997-1999. The nets used in systematic sweeping of the vegetation were made of thick cotton cloth with a diameter of 30 cm at mouth and a bag length of 60 cm. Collections of above ground insects only (from the tall grasses, herbs, shrubs, bushes and trees up to a height of 2 m) were made from each of the study sites, which were divided into 100 quadrates, measuring 10m X 10m each and entire area was covered during sweeping[16]. Sampling from each site was done between 8.00 to 11.00 hr. All of the insects collected from each quadrate were transferred into bottles containing ethyl acetate-soaked cotton. Collected insects were later separated into different taxonomic groups, and a record of different species representing different families of order Coleoptera was maintained. All individuals were oven dried at 60° C for 72 hr. Insects that could not be identified in the laboratory were then identified by scientists of the Entomological Section of Forest Research Institute and Zoological Survey of India, Dehra Dun and Entomological Section of Indian Agricultural Research Institute, New Delhi.

Data Analysis: Diversity (a measure of both the abundance and equitability of species) was expressed using the Shannon - Wiener index (H)[17]. This measure was chosen because it is used extensively in the literature in this context[18-19] and the characteristics of the data set validated its use[17,20,21]. The disadvantages of using H are that the index is a purely relative measure, without absolute meaning[17], but we were particularly interested in comparisons between communities as a function of human disturbance, so the relativity of this index was not problematical.

Having chosen H as the measure of diversity, we used the Shannon-Wiener index of evenness i.e. how the species abundance are distributed among the species[22]. This measure is the ratio of the maximum value of H (assuming that the individuals were evenly distributed among the species) to the realized value of H. As such, this measure of evenness ranges from 0.0 to 1.0. We expressed the species richness as the number of species. Although this measure has the clear advantage of simplicity, it has the disadvantage of over sensitivity and numerically rare species. Findley[23] and Hendrickson and Ehrlich[24] have criticized restricting expressions of diversity to the number of species present while failing to consider the forms and functions of the species. However, our assessment of species richness avoids this pitfall in that we employed concomitant analyses of diversity and we have a reasonable understanding of the ecological functions of the species, which comprise a common feeding guild. Given the sampling constraints, the use of numerical species richness is appropriate[18].

Results and Discussion

Floristic Composition : Ninety-seven species of trees, shrubs and herbs were recorded from all the sampling sites. The site-wise distribution of plant species, recorded in different seasons in all four study sites has been reported elsewhere[25]. The maximum number of plant species (76, 27, 74, 45 in site no. 1, 2, 3 and 4 respectively) were recorded during the rainy season, far fewer were present in the winter season (53, 17, 55, 31) or summer season (47, 21, 56, 31 in site no. 1, 2, 3 & 4 respectively). The greatest numbers of plant species (76) were recorded from site no. 3 (mixed forest with moderate disturbance) which was closely followed by site 1 (mixed forest with no disturbance-75), site 4 (pure sal forest with no disturbance) had 46 species, and only 27 species were recorded from site 2 (plantation forest with severe disturbance).

Table 1. Number of individuals of different species of order coleoptera collected from different study sites of a moist deciduous forest in India during 1997-99.

S. No.	*Taxonomic Group*	*June 97-May98*					*June 98-May 99*				
		Site -1	*Site -2*	*Site -3*	*Site -4*	*Total*	*Site -1*	*Site -2*	*Site -3*	*Site -4*	*Total*
	Cerambycidae										
1.	*Stromatium barbatum* Fabr.	—	1	—	—	1	1	1	—	—	2
	Scarabaeidae										
2.	*Onthophagus pactotlus* Fabr.	2	1	—	—	3	—	2	—	—	2
3.	*Gymnopleurus cyaneus* Fabr.	—	3	—	—	3	4	20	—	—	24
4.	*Mylarbis pustulata* Thunb.	80	4	1	18	103	—	—	—	4	4
5.	*Oxycetonia albopunctata* Fabr.	—	1	—	—	1	3	1	—	—	4
6.	*Ceropria induta* Wied.	1	—	—	1	2	2	—	2	—	4
7.	*Onthophagus* sp.	—	—	1	1	2	1	21	1	—	23
8.	*Citoniinae* sp.	1	—	1	—	2	—	—	1	1	2
	Coccinelidae										
9.	*Coccinella septumpunctata* Linn	14	6	3	10	33	8	16	2	10	36
10.	*Chilomenes sexmaculata* Fabr.	2	2	—	1	5	1	1		—	2
11.	*Rodolia guerinii* Crotch.	22	—	37	21	80	—	—	6	—	6
	Cicindelidae										
12.	*Cicindela aurulenta* Fabr.	1	—	6	—	7	—	2	4	—	6
13.	*Neocollyris crassicomis* Deg.	—	—	2	1	3	—	—	2	1	3
	Lanuidae										
14.	*Monochamus bimaculatus* Gahan	1	1	—	—	2	—	—	—	—	—
15.	*Hoplasama unicolor Tellig.*	14	3	13	47	77	40	4 —	16	58	118
	Curculionidae										
16.	*Myllocerus discolor* Boh.	—	—	—	1	1	—	—	—	—	—
	Total	138	22	64	101	325	60	68	34	74	236

Table 2. Community sturcture of coleoptera in defferent study sites of a moist deciduous forest in India during 1997-1999.

Ecological unit	*Diversity*		*Evenness*		*Richness*		*Abundance*	
	1997-98	*1998-99*	*1997-98*	*1998-99*	*1997-98*	*1998-99*	*1997-98*	*1998-99*
Site 1 (No Disturbance mixed forest)	1.931	1.713	0.581	0.571	10	8	158	97
Site 2 (Severe disturbance . plantation forest)	2.868	2.342	0.905	0.739	9	9	34	78
Site 3 (Moderately disturbance mixed forest)	1.889	2.337	0.630	0.779	8	8	85	64
Site 4 (No disturbance *S.robusta* forest)	2.088	1.061	0.659	0.457	9	5	126	108

Species Composition of Coleoptera

Sixteen species belonging to families Scarabaeidae, Coccinelidae, Chrysomelidae, Cicindelidae, Cerambycidae, Lanuidae and Curculionidae were collected from the study area. In terms of number of species, Scarabaeidae was the most dominant family of Coleoptera, which was represented by seven species, followed by Coccinellidae (3), Cicindelidae (2) and Cerambicidae, Lanuidae, Chrysomelidae and Curculionidae (1-each). Greatest numbers of species (14) were recorded from mixed forest with no disturbance, followed by plantation forest with heavy disturbances (11), pure sal forest with no disturbance (10) and mixed forest with moderate disturbance (9) (Table 1).

On the basis of number of total individuals collected from the study area during 1997-99, Coccinellidae was the most dominant family, constituting 47.78% of all the Coleopterans collected. *Rhodolia guerinii* Crotch was the most dominant species of this family, which constituted 32.08% of all the individuals of this family, followed by *Coccinella septumpunctata* Linn. (25.74%). Chrysomelidae was represented by *Hoplosoma unicolor* Telling, constituting 31.86% of the total Coleopterans. Family Scarabaeidae constituted 36.00% of all the Coleopterans, dominated by species *Mylarbis pustulata* Thunberg (52.97%), followed by *Gymnopleurus cyaneus* (13.36%) and *Onthophagus* sp. (12.37%). Family Cicindellidae constituted 5.35% of all the Coleopterans and families Cerambycidae, Curculionidae and Lanuidae were collected only on a few occasions. A total of 198 insect individuals were recorded from mixed forest with no disturbance (site 1), followed by site 4 (pure *S.robusta* forest-175), site 3 (mixed forest with moderate disturbance-98) and site 2 (plantation forest with severe disturbances-90) during two years of study. The number of Coleopteran individuals was low during February and May but increased during rainy period (July to September) with the increased number of plant species. Such increase in individual numbers was perhaps due to hatching of the eggs in favorable atmospheric temperature and humidity.

Coleoptera Community Structure

During both the study years, the ecological indices of Coleoptera community revealed marked differences among habitats (Table 2). Diversity was maximum in the severely disturbed plantation forest (site 2), followed by undisturbed sal forest (site 4), undisturbed mixed forest (site 1) and moderately disturbed mixed forest (site 3) during the first year of study, while during the second year of study the diversity was recorded maximum in severely disturbed plantation forest (site 2) followed by moderately disturbed mixed forest (site 3), undisturbed mixed forest (site 1) and undisturbed sal forest (site 4).

Evenness also tracked the diversity, with the greatest value being associated with plantation forest, followed by undisturbed sal forest, moderately disturbed plantation forest and undisturbed mixed forest during the first year of study. During the second year of study evenness was greatest in plantation forest, followed by moderately disturbed mixed forest, undisturbed mixed forest and undisturbed sal forest.

The species diversity richness of Coleopteran community showed no marked difference in different study sites during both the study years. During the first year of study richness was ten, nine, eight and nine in sites no. 1, 2, 3 and 4 respectively, while during the second year it was eight, nine eight and five in sites no. 1, 2, 3 and 4 respectively. The greatest abundance of Coleoptera community was recorded from the undisturbed habitats i.e. undisturbed mixed forest (site 1) and undisturbed sal forest (site 4), followed by moderately disturbed mixed forest (site 3) and severely disturbed plantation forest (site 2). The abundance was greatest during the rainy seasons in all the study sites during both the study years except the undisturbed mixed forest (site 1) and moderately disturbed forest (site 3), in which the abundance was greatest in the winter season during the first year of study. The high abundance in site 1 during the winter season of first year was due to the outbreak of Scarabaeidae beetle, *Mylarbis pustulata* Thun., which constituted 92% of the total individuals of order Coleoptera during winter season.

Various workers have also reported species composition of Coleoptera in other ecosystems. Ten species were reported from salt marsh of North Carolina[9], while 169 species from alfa alfa field[10]. Igarshi[11] reported twenty one species at Kawarabi, IBM area, Japan. Janzen and Pond[12] reported 30 and 29 species of Coleoptera at Bradford and Michigan sites, respectively. Vats and Singh[13] reported only 5 species, whereas Kaushal and Vats[14] reporte'd 30 specfies belonging to 7 families from grassland at Kurukshetra, India. Joshi[15] reported 27 species belonging to 7 families from a temperate grassland in india and Joshi and Sharma[26] reported 10 species of Coleoptera representing 6 families from a sub-tropical cropland field.

Species composition and their dominance in any habitat are governed by different factors including the level of disturbance and the presence of teh preferred host plant species. It has been reported by Struhsaker[27] with the help of a flow chart model that maximum diversity remains in the moderately disturbed foret, followed by undisturbed and severely disturbed. During the present investigations, the low diversity of Cleoptera in moderately disturbed site may probably be due to the effect of burining of grasses and cow dung which is practiced by Gujjars not only for cooking purposes but also as an insect repellent. Habitats with only few plant species support relatively fewer species of insects and the favorable effect of rainfall becomes evident when the population of the insects increases with the upsurge of rainfall[28]. Reddy and Alfred[29] reported a highly positive correlation between the total monthly rainfall and the abundance of different groups of chewing and mining insects in Asian forest ecosystems. Fewer numbers of species and individuals recorded in the present study may be due to limiting our investigations to above ground insect faunna only and, these studies of forest insects might under estimate the diversity and abundance of species as well as the complexity of interactions that also occur in forest canopies[30]. However, Corff and Marques[31] reported that species richness was significantly greater by 5-20% in the understory than in the canopy leaves of two dominant tree species, *Querccus alba and* Q. *velutina.* The abundance of the Coleoptera in site 1 may be due to its being a mixed forest with greater number of plant species providing preferred host plants to the insects and with no disturbance. The fewer number of Coleoptera recorded from plantation forest (site 2) may be due to high level of disturbance and also presence of less number of plant species in this site. The present investigations also indicate that in a less diverse forest with higher disturbances, the distribution of species is more even. Thus studying the community structure of Coleoptera may be an important tool in assessing forest ecosystem disturbances and an indicator of prevailing environmental conditions.

Acknowledgement

We are thankful to the Indian Council of Forestry Research and Education (ICFRE) for financial support during the course of this study. Valuable suggestions received from Prof. B. D. Joshi. Dean, Faculty of Life Sciences, GKV, is gratefully acknowledged.

References

1. Resh, V.H. and Barbny, M.A. (1987). *Environ. Entomol.* 16:1087.
2. Danson, W.A. and Travis, J. (1996). *Amer. Enotomol.*, 24:46.
3. Prebble, M.L and Graham, K. (1957). *For. Sci.*, 3:90.
4. Stoszek, K.J. (1973). *J. For.* 71:701.
5. Heikkenin, H.J. (1981). In: The influence of red pine site quality on damage by the European pine shoot moth. *U.S Forest Service Gen. Tech. Rpt. GTR-WO-27.*
6. Beeson, C.F.C. (1941). In : *The ecology and control of the forest insects of India and the neighboring countries.* Vasant Press, Dehradun, 1.
7. Roonwal, M.L. (1954). *Indian For. Bull.* 171:(1).
8. Mathur, R.N. and Singh, B. (1959). *Indian For. Bull.* 171: 3.
9. Davis, L.V. and Gray, I.E. (1966). *Ecol. Monogr.* 36 (3):275.
10. Evans, F.C. and Murdoch, W.W. (1968). Taxonomic composition, trophic structure and seasonal occurrence in a grasland insect community. In *A Grassland Insect Community.* 37:259.
11. Igarshi, R. (1973). Arthropod fauna. In: *Ecological studies in Japanese Grassland with special reference to IBM area: productivity of terrestrial communities* (M.Numata Ed.), Japanese IBM synthesis (Tokyo: University of Tokyo Press). 13:227-237.
12. Janzen, D.H. and Pond, C.M. (1975). *Trans. R.Entomol. Soc.* London 127:33.
13. Vats, L.K. and Singh, J.S. (1978). *Tropical Ecology.* 19(1): 51.

14. Kaushal, B.R. and Vats, L.K. (1987). *Entomon.* 12 (2):161.
15. Joshi, P.C. (1996). *Uttar Pradesh J.* Zool. 16(3): 169.
16. Gadagakar, R., Chandrashekhar, K. and Nair, P. (1990). *J. Bom. Nat. Hist. Soc.*, 87 (3):328.
17. Southwood, T.R.E. (1978). In:*Ecological methods.* Chapman and Hall, New York.
18. Magurran, A.E. (1988). In: *Ecological diversity and its measurement.* Princeton University Press, Princeton.
19. Spellberg, I.F. (1992). In: *Evaluation and assessment of conservation.* Chapman and Hall, New York.
20. May, R.M. (1975). Patterns of species abundance and diversity. In : *Ecology and evolution of communities* (M.L. Cody, J.M. Dimond, eds.). Harverd University Press, Cambridge, M.A. p. 81
21. Taylor, L.R. (1987). *Biological Conservation.* 39:209-234.
22. Kotlia, P.M. (1986). Ecological measures. Software package, St. Lawerance University.
23. Findley, J.S. (1973). *American Naturalist.* 107:580.
24. Hendrickson, J.R. and Ehrilch, P.R. (1971). An expended concept of species diversity. *Notulae Naturae.* Academy of Natural Sciences of Philadelphia 439:1-6
25. Joshi, P.C., Lockwood, J.A., Vashishth, N. and Singh, A. (1999). *J. Orthoptera Research.* 8:17.
26. Joshi, P.C. and Sharma, R. (1998). *Uttar Pradesh J.* Zool. 18(1):1.
27. Struhsaker, T.T. (1987), *Biological Conservation.* 39:209.
28. Trojan, P. (1975). *Ekologia ogolina.* Polish Acad. Sci., Warsaw.
29. Reddy, M., Vikram, J.R. and Alfred, B. (1977). In: An inexpensive easily portable light trap for taxonomical and ecological studies on forest insects. *Second Oriental Entomol. Sym.* p. 48.
30. Lowman, M.D. and Wittman, P.K. (1996). *Ann. Rev. Ecol. Syst.* 27:55.
31. Corff, J.L. and Marquis, R.J. (1999). *Ecol. Enotomol.* 24:46.

2

Biodiversity and Abundance of Hemiptera, Coleoptera and Water Characteristics of Lake Mansar

BALDEV SHARMA AND SHALOO AYRI
Department of Zoology University of Jammu, Jammu-180006

Abstract

Lake Mansar is a perennial natural aquatic resource covering an area of 3.29 sq kms and is a famous destination in Jammu. Consequently the Lake is vulnerable to caused pollution by the sewage and fertilizers. So the present investigation was carried out as a consequence of which 9 species of Hemiptera and 8 species of Coleoptera have been reported, some of which are the indicators of oligotrophy while others are the indicator of eutrophy, so the lake seems to be undergoing Eutrophication.

Introduction

The study of aquatic insects in India has been carried out from time to time. But still various aquatic resources are there which are untouched and unexplored. The Jammu and Kashmir state has still most of the water bodies virtually virgin regarding the study of aquatic insects. Salaria[1] consolidated the list of aquatic insects reported in J&K State and recorded 98 species of aquatic insects. Thakur[2] surveyed on Hemiptera and Coleoptera inhabiting freshwater aquatic water bodies around Trikuta Hills, J&K. Present paper focuses on certain species of aquatic hemipterans and coleopterans present in this lake and which have indicated that the lake is undergoing Eutrophication.

Materials and Methods

Lake Mansar is situated at an elevation of 666 meters of msl in Shivalik range about 65 kms east of Jammu City and located between 32° 45" North and 75° 23" East. Mansar Lake encloses an area of 329.4 ha with a maximum depth of 37.6 meters at the center. Lake has subtropical monsoonal climate and experiences monsoon rains from July to September. The average rainfall is 1500 mm. The temperature during summer ranges between 35° C and 40° C. Aquatic insects were collected fortnightly by using following methods:

1. Net sweeping. In deep water insects were collected by sweeping net through it.
2. By examining stones. Most of the insects were found below stones and thus manually trapped.
3. By hand picking. Skillful hand picking can easily collect large sized insects.

Results and Discussion

Lake being a famous tourist destination is vulnerable to third category of pollution besides the sewage and fertilizers. In addition to this, lake is also revered for its religious sanctity. Devotees come here for performing various religious rights and rituals. They take holy dip in the water; wash their clothes using detergents and conduct *mundane* ceremonies for which functions are organized with ground community participation. Presence of a zoo and a cremation ground are a regular source of pollution, as runoff water add outcomes to these sites into the lake and threatens the lake ecology leading to eutrophication. During present study of Lake Mansar following species were reported at different stations. Following Table 1 shows the distribution of reported species at different stations showing abundance of Hemiptera and Coleoptera at two different sites.

At the sites like zoo and cremation ground there is absence of any sort of vegetation. They are barren and blunt peripheries of lake without any vegetation. But some sites are there which are with rich vegetation and show the presence of macrophytes viz. *Nymphoides indicum, Arundo donex, Typha augustata, Nelumbium pentaplata*[3]. The habitat difference provided by these stations to the insects shows the great effect on presence and absence of above reported species.

Species *Heleocoris* belonging to family Naucoridae has been reported as oligotrophic indicator[4] and majority of Hemipterans belonging to family Nepidae, Belostomatidae, Gerridae from vegetative sites represent oligotrophic type of status of these non polluted sites.

Table 1. Abundance of Insects of different groups

Insects Vegetative sites	*Zoo & Cremation sites*	*Rich*
Order: Hemiptera		
Heleocoris sp.	–	+ + +
Gerris sp.	+	+ + + +
Plea buenoi	+	+ + + +
Paraplea buenoi	+	+ + +
Sphaeroderma molestrum	+	+ + +
Micronecta sp.	+	+ + + +
Laccotrephus maculates	+ +	+ + + +
Ranatra filliformes	+ +	+ + + +
Order: Coleoptera		
Laccophilus sp.	+ + + + +	–
Cybister tripuctatus	+ + + +	–
Amblystomus erichson	+ + + + +	–
Canthydrus sp.	+ + + +	–
Sternolophus rufipus	+	+ + + +
Bagous sp.	+	+ + + +
Berosus sp.	+ + + +	+ +
Dactylosternum	+ +	–
Bidessus sp.	+ +	–

On the other hand species which act as eutrophic indicators e.g. family Dytiscidae representing species like *Cybister tripunctatus, Laccophilus leach, Canthydrus leatibllls* Walker, *Didessus* sp. are collected mainly from zoo and cremation sites. This indicates that undesirable things are being added to the lake in these stations leading to the eutrophication of this lake. These sites show complete absence of oligotrophic indicators like *Heleocoris* sp., members of Hydrophilidae family like *Sternolophus* sp., *Berossus* sp., *Dactylosternum* sp., *Enochrus* sp. though they have been recorded from stations with rich macrophytes.

Presence of both types of indicators i.e. wide range tolerant ones and narrow range tolerant from the same water body but from different microhabitats provided by surrounding lake conditions indicates that the lake is acquiring eutrophic status day by day although water is still clear enough thereby indicating declining oligotrophic status.

Acknowledgement

Authors are thankful to Dr. V.V. Ramamurthy (IARI) for his help in identification of reported species.

References

1. Salaria, R. (1992). In: *Taxonomic survey of aquatic Hemipterans of Jammu Region J&K state. M. Phil. Dissertation,* University of Jammu, Jammu.
2. Thakur, S. (2003). In: *Survey of Coleoptera and Hemiptera inhabiting water bodies around Trikuta Hills. M. Phil. Dissertation,* University of Jammu, Jammu.
3. Sharma, S. (1991). In: *Contribution to the aquatic and marshy flora of Jammu.* Ph.D.Thesis, University of Jammu, Jammu.
4. Bisht, R.S. and Das, S.M. (1979). *Ind. Jour. Ecol.* 6(1): 35.

3

Diversity, Distribution and Relative Abundance of Insect Visitors on *Brassica Compestris* Var. Sarson (J&K)

J.S.TARA AND POOJA SHARMA
Department of Zoology, University of Jammu (J & K) - 180 006

Abstract

Studies on insect diversity showed that flowers of the plant *Brassica compestris* were visited by 36 species of insects belonging to 5 orders and 15 families of class Insecta at two field stations. Of these, 15 species belonged to order Hymenoptera, 13 to Diptera, 05 to Lepidoptera, 01 to Coleoptera and 02 to Hemiptera. Apis dorsata was the most abundant pollinator visiting the plants blooms with their mean number as 45.61±0.25, (39.30%) at Pallimore and 63.82±0.25 (49.05%) at Hiranagar fields stations, in district Kathua followed by Apis mellifera 30.93+0.16, (26.65%) and 28.22±0.22 (21.69%) and Apis cerana 10.97±0.15 (9.45%) & 19.68±0.06 (15.13%) respectively at Pallimore and Hiranagar fields. Their mean abudance was significantly higher (P< 0.01) than other insect pollinators.

Introduction

Brassica compestris var. Sarson is a self-compatible oilseed crop belonging to family Brassicaceae and is also cross-pollinated by number of insects particularly the honeybees for producing more seeds of better quality[1,2]. Pollination of entomophilies crops by bees is one of the most effective and cheapest methods of increasing the yield. Other agronomic practices like manuring, pesticides, irrigation and fertilizers are also important but are cost effective and may not yield desired results without the use of honeybees as pollinators[3]. Honeybees are considered as the most efficient pollinators of cultivated crops because of their floral fidelity[4,5] , potential for long working hours[6], presence of pollen baskets, maintainability of high population, micromanipulation of flowers and adaptability to different climatic conditions[2].

Materials and Methods

The studies on diversity, distribution and relative abundance of insect visitors on *Brassica compestris* var. Sarson were carried out at two field stations - Pallimore and Hiranagar, of district Kathua, J&K in 2006, which are summarized as under.

(1) Diversity of Insect Visitors on Sarson : Through out the blooming period, insect visitors were collected by polythene bags, dried and preserved by traditional methods. Identification of the preserved specimens was done from the taxonomy section of "Entomology Division" of Indian Agricultural Research Institute (IARI), New Delhi.

(2) Relative Abundance of Insect Pollinators : Relative abundance of insect pollinators was recorded in terms of their visits in a quadrate of 1x1m size[7] for 5 minutes. The observations were repeated at two hours intervals starting from 1000-1600 hour. This gives the abundance of insect visitors for that particular day. The observations were repeated for 7 days in each field. Temperature and relative humidity at the time of each observation were also recorded with the dry and wet bulb thermometer.

Results and Discussion

In total 36 species of insects belonging to 5 orders and 15 families of class Insecta were reported to visit the sarson blooms, at both the field stations (Table 1). Of these, 15 species belonged to order Hymenoptera, 13 to Diptera, 05 to Lepidoptera, 01 to Coleoptera and 02 to Hemiptera. However, Rahman [8] reported 105 different species of insect pollinators belonging to 55 families and 9 orders of the class Insecta visiting toria and sarson bloom at Lyallpur where as, Bhalla *et, al.* [9] recorded 7 species of insect visitors on mustard bloom at Solan (HP). The op. *cited* authors also observed honeybees and dipteran flies as the predominant visitors and worked for longer hours throughout the flowering season. Some similar findings were also made by the present author where, Hymenopterans constituted 87.96 and 93.44 percent of total insect visitors followed by Dipterans, 10.18 and 4.65%; Lepidopterans, 0.34 and 0.41% and others* as 1.53 and 1.51% respectively at Pallimore and Hiranagar fields (Table 1).

With regards to the relative abundance, mean number and percentage of the large rock bee, *Apis dorsata* was found to be highly abundant i.e., 45.61 ±0.25/m^2, 39.30% and 63.82±0.25/m^2, 49.05%, followed by the exotic bee, *Apis mellifera* (30.93±0.16/m^2, 26.65% & 28.22+0.28/m^2, 21.69%) and the Indian hive bee, *Apis cerana* (10.97±0.15/m^2, 9.45% & 19.68±0.06/m^2, 15.13%) at both the fields. The percent population of *Apis dorsata, A. mellifera* and *A. cerana* was significantly higher ($P< 0.01$) than other insect pollinators *viz., Apis florea, Andrena* sp., *Eristalis tenax* , *Eristalis* spp., *E.solitus, Syrphus* sp., *Eumerus* sp., *Musca domestica. Orthellia* sp., *Pieris brassicae* and others*(*Coccinella* sp., 04 species whose taxonomic status not determined- 02 to Hymenoptera, 01 to Diptera, 01 to Hemiptera)(Table 2 & 3). Similarly, Sharma et al.[10] observed Apis dorsata as the best pollinator of sarson followed by *A. mellifera* and *A. florea* at Hissar, Haryana. Again, Chaudhary [11] reported *A. dorsata* population almost exclusively on rape cultivars. But maximum population of *Apis cerana indica* was reported to visit mustard bloom at Pusa [12] followed by *A. dorsata, A. mellifera* and *A. florea* bees and other insects of the order Hymenoptera, Lepidoptera and Diptera. Similar observations were reported on rapeseed bloom at Palampur[13], on sarson, at Nauni[14], on brown mustard in Bihar[15]. However, Singh et, al.[16], reported *A. florea* outnumbering (23.65) *A. mellifera*, while *A. dorsata* population was least abundant (0.09) on toria at Pusa (Bihar). Further, it has been observed that maximum activities of insect pollinators remained between 1000-1600 hours thereafter, their population declines. Earlier similar observation was made by Rahman[8] on toria and sarson bloom at Lyallpur. This study permits us to infer that *Apis dorsata* was the efficient pollinator of sarson bloom followed by *Apis mellifera* and *Apis cerana* in uplifting the ongoing "yellow and white revolution" consequent to "green revolution".

Table 1. Diversity .distribution and relative abundance of insect visitors

Older Family	*Genus Species*	*% pop. Pallimore*	*% order*	*% pop. Hiranagar*	*% order*
Hymenoptera/					
Apidae	*Apis cerana* (PH)	9.45		15.13	
	Apis mellifera (PH)	26.65		21.69	
	Apis dorsata (PH)	39.30		49.05	
	Apis florea (PH)	3.54		2.33	
	Bombus sp.(P)	+		–	
Andrenidae	*Andrena* sp.(PH)	9.02		5.42	
Vespidae	*Eumenes edwardsii* (H)	-		+	
	Vespa basalis (P)	+	87.96	-	93.44
	Rhynchium pitidulum (P)	+		-	
Megachilidae	*Megachile* sp.(H)	-		+	
Halictidae	*Halictus* sp.(H)	-		+	
Anthophoridae	Xylocopa aestuans ♀ (H)	-		+	
	Xylocopa aestuans ♂ (H)	-		+	
Diptera/					
Syrphidae	*Eristalis tenax* (PH)	1.94		0.91	
	Eristalis spp. (PH)	3.60		-	
	Eristalis solitus (PH)	1.95		-	Contd.
	Syrphus sp. (PH)	1.38		0.52	
	Eumerus sp. (PH)	0.74		-	
	Xanthogramma citrina (P)	+		-	
	Syritta sp. (H)	-		+	
	Eristalis quinquelineatus (H)	-		+	
					4.65
Muscidae	*Orthellia* sp. (PH)	+		0.65	
	Musca *domestica* (PH)	0.57		+	
	Orthellia caeserion (H)	-		+	
Tephritidae	*Bactrocera dorsalis* (H)			+	
Lepidoptera/					
Pieridae	Pieris *brassicae* (PH)	0.34		0.41	
	Delias sp. (P)	+			
Lycaenidae	*Deudorix* sp.(P)	+			
Syntpmidae	Syntomis *sperbius* (P)	+	0.34	-	0.41
Nymphalidae	Precis almana (H)	-		+	
Others					
Coleoptera/					
Coccinellidae	*Coccinella* sp. (PH)	+		+	
*Hymenoptera					
	Taxonomic status not determined (PH)	1.53	1.53	1.51	1.51
*Diptera					
Hemiptera/					
Pentatomidae	Eurydema sp. (P)	+		-	

P= Pallimore; H= Hiranagar; + = Present; - = Absent (Occassional visitors)
Species with their % at both Pallimore and Hiranagar were the frequent visitors

Table 2. Mean abundance of insect pollinators visiting sarson bloom at Paalimore (No. of insects/m²/5 minutes) at different hours of the day.

Order/ Family	Genus/ Species	*Different hours of the day* Mean S.E. 1000 hr	MeantS.E. 1200 hr	Mean+S.E. 1400 hr	Mean+S.E. 1600 hr	µ±S.E.	% population	% Order
Hymenoptera/ Apidae	*Apis cerana*	6.57±1.11	16.14±0.70	14.29+0.71	6.86+0.40	10.97±0.15	9.45*	
	Apis mellifer	30.57+1.74	37.43±2.27	32.57+1.99	23.14±1.52	30.93±0.16	26.65*	
	Apis dorsata	47.14±2.82	54.00±2.09	54.57±1.84	26.71±1.66	45.61±0.25	39 30*	
	Apis florea	1.29–0.33	2.71±0.47	7.7U0.47	4.71±0.86	4.11+0.11	3.54	87.96 .
Andrenidae	*Andrena* sp.	6.71+0.61	9.29+1.02	17.29±1.64	8.57±0.20	10.47±0.31	9.02	
Diptera/ Syrphidae	*Eristalis tenax*	2.57+0.48	2.86+0.40	1.71+0.29	1.86+0.46	2.25±0.04	1.94	
	Eristalis spp	5.00±0.95	5.00+1.41	5.00+0.82	1.71±0.42	4.18+0.20	3.60	
	Eristalis solitus	3.43+0.37	2.44+0.40	1.43+0.20	2.00±0.38	2.25±0.05	1.95	
	Syrphus sp.	1.57+0.48	2.14+0.34	1.71+0.61	1.00+0.38	1.60+0.06	1.38	10.18
	Eumerus sp	1.43+0.45	1.00+0.00	0.43+0.30	0.86+0.55	0.86±0.12	0.74	
Muscidae	*Musca domestica.*	1.43+0.42	0.43±0.20	0.29+0.18	0.51±0.30	0.66+0.05	0.57	
Lepidoptera/ Pieridae	*Pieris brassicae*	0.00±0.00	0.86±0.26	0.43+0.20	0.29+0.18	0.39+0.06	0.34	0.34
Others								
(Coleoptera, Hymenoptera, Diptera, Hemiptera)	Taxonomic status not determined except *Coccinella* sp.	1.71±0.36	2.43+078	1.43+0.48	1.57±0.37	1.79±0.10	1.53	1.53

±S.E = Standard error about mean; MeaniS.E of 35 observations; µ±S.E of 140 observations; *= Mean abundance of *Apis dorsata, A. mellifera* and *A. cerana* was higer than other insect visitors (P<0.01).

Table 3. Mean abundance of insect pollinators visiting sarson bloom at Hiranagar (No. of insects/m²/5 minutes) at different hours of the day.

		Different hours of the day						
Order/ Family	Genus/ Species	MeaniS.E. 1000 hr	MeantS.E. 1200 hr	Mean+S.E. 1400 hr	Mean+S.E. 1600 hr	μ±S.E.	% population	% Order
Hymenoptera/								
Apidae	*Apis cerana*	22.29±0.68	27.0010.39	17.14±0.51	12.29±0.52	19.68±0.06	15.13*	
	Apis mellifera	30.29±1.06	30.86±1.83	27.71±1.02	24.00±0.49	28.22+0.28	21 .69*	
	Apis dorsata	63.14±1.37	63.57±0.68	65.00±1.45	63.57±1.90	63.82±0.25	49.05*	93.44
	Apis florea	2.86±0.40	3.43±0.43	3.71±0.47	2.14±0.46	3.03±0.01	2.33	
Andrenidae	*Andrena* sp.	4.86±0.34	7.29±0.42	8.86±0.34	6.29±0.29	6.82±0.03	5.24	
Diptera/								
Syrphidae	*Eristalis tenax*	1.86±0.34	1.71±0.18	0.71±0.29	0.43±0.30	1.18±0.03	0.91	
	Eristalis spp.	2.57±0.37	2.34±0.37	2.00±0.31	1.29±0.36	2.07±0.01	1.59	
	Eristalis solitus	2.71±0.52	1.00±0.22	0.29+0.18	1.14±0.34	1.28±0.08	0.98	4.65
	Syrphus sp.	1.57+0.20	0.43±0.20	0.57±0.37	0.14±0.14	0.68±0.05	0.52	
Muscidae	*Orthellia* sp.	1.29±0.36	0.43±0.20	0.00±0.00	1.71±0.29	0.68±0.08	0.65	
Lepidoptera/								
Pieridae	*Pieris brassicae*	0.43±0.30	0.57±0.20	0.57±0.20	0.57±0.20	0.53±0.02	0.41	0.41
Others **(Coleoptera, Hymenoptera, Diptera, Hemiptera)**	Taxonomic status not determined except *Coccinella* sp.	1.71±0.68	1.57±0.20	1.86±0.55	2.71±0.47	1.96±0.10	1.51	1.51

±S.E = Standard error about the mean; Mean±S.E of 35 observations; μ±S.E of 140 bservations. * = Mean abundance of *Apis dorsata*. *A. mellifera* and *A. cerana* was higher than other insect visitors (P< 0.01).

Acknowledgment

The authors are thankful to the Head, Department of Zoology, University of Jammu, Jammu for providing research facilities. The authors are indebted to Dr. V.V.Ramamurthy, Principal Scientist, Division of Entomology, IARI, New Delhi for Identification of Insects. Thanks are also due to Dr. V.K.Mattu, Department of Bio-Sciences, H.P University, Shimla for the encouragement and guidance.

References

1. Free, J.B. (1993). In : Insect Pollination of Crops. Academic Press, New York, p. 864.
2. Verma, L.R. and U. Partap (1993). In : The Asian Hive Bee, Apis cerana, as a Pollinator in Vegetable Seed Production (An Awareness Handbook). International Centre for Integrated Mountain Development (ICIMOD), Kathmandu, Nepal.
3. Verma, L.R. (1990). In: Beekeeping in Integrated Mountain Development: Economic and Scientific Perspective. Oxford and IBH Publ. Co. Ltd. New Delhi.
4. Wells, H. and P.H. Wells. (1983). *J. Anim.Ecol.* 52: 829.
5. Waser, N.M. (1986). *Amer. Nat.* 127 (5): 593.
6. Sihag, R.C. (1990). *Indian Bee J.* 52 (1-4): 51.
7. Priti and R.C. Sihag. (1997). *Indian Bee J.* 59 (4): 230.
8. Rahman, K.A. (1940). *Indian J. Agri. Sci.* 10: 422.
9. Bhalla, O.P., A.K. Verma and H.S. Dahlia. (1983). *J. Ent. Res.* 7 (1): 15.
10. Sharma, S.K., J.R. Singh and Ombir (2000). Crop-Research- Hissar, 19 (1): 125.
11. Chaudhary, O.P. (2001). *Insect Environment.* 7 (3): 141.
12. Chand, H., R. Singh and S.F. Hameed (1994). *J. Ent. Res.* 18(3): 233.
13. Rana, V.K., R. Desh and R. Kaushik (1997). *J. Ent. Res.* 21 (1): 59.
14. Mishra, R.C., J. Kumar and J.K. Gupta (1988). *J. Apic. Res.* 27 (3): 186.
15. Prashad, D., S.F. Hameed, R. Singh, S.S. Yazdani and B. Singh. (1989). *Indian Bee J.* 51 (2): 45.
16. Singh, B., M. Kumar, A.K. Sharma and L.P. Yadav. (2004). *Environment and Ecology.* 23 (3): 571.

4

Diversity and Richness of Butterflies: A Case Study in A High Altitude Forest in Uttarakhand, India

P.C. JOSHI
Department of Zoology and Environmental Science, Gurukula Kangri University, Hardwar - 249404, India

Abstract

Species Composition of Butterflies was studied in Pindari area of Nanda Devi Biosphere Reserve in Bageshwar district of Uttarakhand state in India. Study area supports sub-alpine and alpine type of vegetation along its entire altitudinal range. Collections were made from three sites covering a total of 88 Km^2 and located at different altitudes with different degree of anthropogenic disturbances. A total of 3692 individuals belonging to 46 species and 15 families of order Lepidoptera were recorded. Family Nymphalidae with 10 Species and 1091 individuals was the most dominant family. It was followed by family Pieridae with 9 species and 958 individual and Lycaenidae with 06 species and 497 individuals. *Vanessa cashmirensis* Fru. (Nymphalidae) was the most abundant species present in all the three study areas and constituting 8.1% towards total number of individuals collected from all the three sites. It was followed by *Pieris canidia indica* sparr. (pieridae) (6.3%) and *Heliphorous sena* (Lycaenidae) (4.8%). Site 1 which was at the lowest altitude supported as many as 42 species and 1625 individuals representing 91.3% and 44% of total number of species and individuals, respectively recorded from the Pindari area. In this study site *Vanessa cashmirensis* Fru was the most dominant species constituting 10.1% of total individuals. Eight species of Lepidoptera were confined to this site (Site 1) only, which included *Atella phalanta phalanta* Drury,

Anaphaeis aurota (Fabr.), *Huphina herissa phryne* (Fabr.), *Chilaria kina* (Hewiston), *Papilio polytes romulus Cramer.*, *Danaus aglae* Stoll., *Macroglossum* sp., *Agylla rufifrons* Moore. Six species of lepidopteran insects were confined to sites 1 and 2 which included *Pseudoergolis wedah* Kollar, *Sephisa dichroa* Kollar, *Rahinda Hardonia* Stoll, *Euploea core* (Cramer), *Calpe ophideroides* Guen, *Gazalina chrysolopha* Koll. And *Estigmene so.*, while three species were recorded from area 2 and 3 only viz. *Argynmis lathonia issea* Doubleday, *Calias erate* (Esper.) and *Chalesia pectinicornis albata* Moore. *Panaera mydon* Walk, was confined to site 3 at the highest altitude. The least abundant species recorded during the present study included *Atlla phalanta phalanata* Drury, *Chilaria kina* (Hewiston) and *Panaera mydon* walk. The community structure of butterflies further showed a clear influence of ecological conditions associated with altitude and also with the disturbance in the area as the highest value of diversity (3.325), evenness (0.751) and richness (42) were recorded for the site situated at lowest altitude with moderate level of disturbances.

Introduction

Insects play a vital role in maintaining the cycling of nutrients, soil regeneration and protection, pollination of phanerogamic plants as well as natural regulation of pest out breaks. Among insects, the butterflies are ecologically very important. The adults generally feed on nectar and are important as pollinators of flowering plants. The larvae, which feed on foliage, are the primary herbivores in the ecosystem and are important in the transfer of radiant energy fixed by plants and making it available to the other organisms in the ecosystem. Lepidoptera, specially the butterflies, show distinct pattern of habitat association. The nature of vegetation, humidity, sunshine, availability of water etc. are factors that determine the survival of given species in a particular habitat. The order Lepidoptera which contains the butterflies and moths is the second largest insect order with 1,40,000 species reported from all over the world of which 12,000 are butter flies. They have great aesthetic value and many species are much sought after because of their marvelous appearance. The immature and adult stages of butterflies depend entirely on plants and hence are economically and ecologically important. They are important pollinators of several wild and domesticated plant species and depletion of their population could adversely affect the regeneration of plants they pollinate.

All stages of butterflies are also fed upon by various higher groups of animals like birds, bats and mammals and thus form more than one link in the food web. The presence of butterflies emphasizes availability of the larval food plants in great abundance. Most butterflies have specific habitat requirements, female usually tend to lay eggs only on selective food plants occurring in the area. Although adult butterflies are also selective in their choice of the flowers for feeding, most species never feed on same flowers as they become available. There is an intimate association between butterflies and their food plants.

Butterflies serve as good indicator of the environmental changes that occur as humans develop the landscape and as excellent indicator of urbanization. Many researches have suggested that butterflies are suited as indirect measures of environmental variation because they are sensitive to local weather, climate and light levels. Further, butterflies diversity may serve as a surrogate for plant diversity because butterflies are directly dependent on plants often in highly coevolved situation. More over, few studies have attempted to relate changes in insect communities to the degree of alteration of the forest habitats resulting from disturbance. This is particularly unfortunate because many insects respond to disturbance more rapidly than do vertebrates, giving them potential as early indicators of ecological change.

Both the habitat preferences and the geographical ranges of species are closely related to their life-history strategies and this suggests that disturbance would be expected to disproportionately affect those species with narrow geographical ranges. Diversity can be a powerful comparative index when used to determine the relative state of a system after human disturbance. In tropical rain forest for example, the impact of logging and cultivation and subsequent recovery from these activities could be measured in terms of faunal diversity. Ecosystem disruption through human activity is judged to pose a major threat to the long term prospects of biodiversity conservation. As insects form a significant proportion of global species diversity the future of the earth's biological resources is very much tied up with the conservation of insect diversity. The present study carried out in a highly sensitive ecological zone of Himalayan region, describes the role of different ecological factors on distribution of butterflies.

Materials and Methods

Study Area

The present study has been carried out in the Pindari area located (30° 5'30° 10' N to 79° 48'-79° 52'E) in the northern part of Bageshwar district of Uttaranchal. This area lies at buffer zone of Nanda Devi

Biosphere Reserve in Western Himalayas. The area is connected with road head by a footpath and boundary of the buffer zone starts after traveling 20 km from the road head. The main river is Pindar, which is joined by its tributaries Sunderdhunga and Kaphni rivers in the buffer zone. The Pinder river originates from the Pindari glacier. The entire area represents a characteristic high altitude environ and falls within an altitudinal range of 2100- 7000m. The glaciers in the area viz. Pindari, Kaphni and Sunderdhunga are main attractions for mountain and ecotourism.

Climate

Climatically, the area is unique and has three seasons i.e. winter (November to March), summer (April to mid June) and rainy (mid June to mid October).

Vegetation

Due to altitudinal and climatic gradients, the natural vegetation changes from temperate to sub-alpine and alpine type. Broad-leaved deciduous and evergreen coniferous forests mainly characterize the temperate and sub-alpine vegetation. Herbs, sedges, grasses and few scattered shrubs mainly dominate the alpine vegetation.

Disturbances

The Pindari area, named after a famous glacier in the area, has witnessed a tremendous increase in the human population, which have exploited the natural flora and fauna. It has also witnessed some geological activities over the past few decades. Thus the disturbances in the study area are either anthropogenic or geological. The anthropogenic disturbances include tree cutting, lopping of trees (cutting of smaller branches full of green foliage) for fodder and fuel wood, removal of leaf litter from the forest floor and surface burning due to fire. Land slides and soil erosion mainly contributed geological disturbances, which lead to the removal of top soil and nutrients. Succession takes place after the disturbances and is arrested by the recurrent anthropogenic or geological disturbances.

During the present study collections were made from three sites located in the buffer zone of Nanda devi Biosphere Reserve. These three areas represented different altitudes, vegetation type and level of disturbances and were located within a distance of 22 kms from each other in the buffer zone. Table 1 shows the characteristic features of all the three study sites.

Site No. 1 (Khati)

Khati is connected with road head by a footpath after traveling 22 km from Song village (Bageshwar district) and is close to the river Pindar. It is located at an altitude of 2100 m. The forest is more of a Pangur *(Aesculas indica* Colebr. Ex. camb) Hk., which is associated with dominant forest of *Quercus floribunda* Rehder, *Junglens regia* L, *Aluns nepalensis* Don., *Acer villiosum* Wall, and *Cedrus deodara* (Roxb.) Loud. This site is moderately disturbed because of its proximity to Khati village and receives moderate disturbances due to the grazing by cattle and collection of forest products by the neighboring villagers. During the study period temperature of this study site varied between 9°C to 26°C, while the relative humidity ranged between 30% to 75%.

Site No. 2 (Dawali)

The distance between Khati and Dawali is 12 km. This site is located at an altitude of 2800 m and is dominated by colorful Burans *(Rhododendron arboreum* Sm. and *R. barbatum* Wall.) forest associated with mixed forest of *Acer cappadocium* Gled, *Taxus baccatata* L. subsp., *Acer caesium* Wall. ex. Brand, *Abies pindrow* Royle and *Ulmus wallichiana* Plank. This site has very low level of anthropogenic disturbances. The V-shaped valley (both side of Dawali) is drained by river Pindar and very sensitive to geological disturbance i.e. land slides, rock fall and soil-erosion activities. Temperature at Dawali ranged between 6°C to 22°C, while relative humidity ranged between 35% to 80%.

Site No. 3 (Phurkia)

This site is located at an altitude of 3500m and is close to the Pindari glacier and core zone of Nanda Devi Biosphere Reserve. The distance between Dawali and Phurkia is 7 km. This study area represents to a moist sub alpine scrub and alpine forest. The dominant plant species include Bhoj *(Betula utalis* D.Don). *Acer acuminatum* Wall., *Abies spectablis* (D.Don) Mirle), *Rhododendron campanulatum* D.Don and R. *anthopogon* D.Don. This is a highly snow prone area of buffer zone which starts receiving snow from last weak of November, which can be seen there till the end of March. This site is highly disturbed due to altitudinal migration of tribes who use the area as their summer settlement and collect fuel and fodder apart from the area being used for grazing by their cattle and sheep. Temperature at this site ranged between 4°C to 19°C and relative humidity ranged between 40% to 90%.

Sampling of Entomofauna

The insects were collected by sweep sampling method. The net sweeps were carried to collect the insects. The nets used in systematic sweeping were made of thick cotton cloth with a diameter of 30 cm at mouth and a beg length of 60 cm. To assure a consistent, systematic sampling, a randomly selected area of each study site was divided into 100 quadrates of 10 × 10m. Sampling was done at random and at an interval of 30 days. The collected insects were transferred into jars containing Ethyl Acetate soaked cotton. These jars were brought to the laboratory and the insects were stretched and pinned. The entomological pin numbers 1 to 20 were used according to the size of the specimen. These were oven dried at 60°C for 72 hours to preserve them and then set in to wooden boxes and labeled according to their systematic position. The species, which could not be identified in the laboratory, were sent to Northern Regional Station of Zoological Survey of India, Dehradun, Entomological Section of Forest Research Institute, Dehradun, Indian Agricultural Research Institute, Pusa, New Delhi and Zoological Survey of India, Kolkata for their identification.

Data Analysis

The diversity of butterflies was expressed using Shannon-Wiener Index[1] (H). For determining the evenness, Shannon-Wiener index of evenness was used[2], which ranges from 0.0 to 1.0.

Results

Floristic Composition

During our study a total of 144 plant species, including 26 trees, 45 shrubs and 73 herbs were recorded from all the three areas surveyed. The site at lowest altitude supported maximum number (81) of plant species, which included 15 species of trees, 26 species of shrubs and 40 species of herbs and grasses. It is a temperate forest characterized by broad-leaved deciduous and evergreen forest of *Aesculus indica* (Cobler. Ex Don) Hk., *Junglens regia* Lin., *Quercus floribunda* Reheder, *Alnus napalensis* Don. *Acer villiousum* Wall., and *Cedrus deodara* (Roxb.) Loud.

The second study site supported as many as 78 species of plants including 13 species of trees, 24 species of shrubs and 41 species of grasses and herbs. The dominant tree species included *Acer cappadocium* Gled, *Rhododendron barbatum* Wall., *R: arboreum* Sm., *Taxus baccata L* and *Ulmus wallichiana* Plank. The third study site at highest altitude represented a moist sub-alpine scrub and alpine forest and supported 63 species of plants with 8 species of trees, 17 species of shrubs and 38 species of other grasses and herbs. *Betula utalis D.* Don, *Salix disperma* Roxb., *Abies spectablis* (D. Don) Mirle and *Prunus cgrnuta* (Royle) Steud. were the dominant tree species.

Table 1. Characteristic features of different study sites selected in the present study during 2002-04.

Study sites	*Altitude*	*Ecological Zone*	*Dominant plant community*	*Disturbenses*	
				Anthropogenic	*Geological*
Site no. 1 (Khati)	2100m	Temperate	*Quercus floribunda* Render.*Juglens regia L. Acer villiosum* Wall., *Aeculus indica* (Colebr. Ex. Camb) Hk.	Moderate level	land slides hill slides
Site no. 2 (Dawali)	2800m	Sub-alpine	*Rhodendron barbatum Wall. Acer cappadocium* Gled. *A. caesium* Wall. Ex. Brand	Low level	Erosion, Rock fall
Site no. 3 (Phurkia)	3500 m	Alpine	*Betula utalis* D.Don, *Acer acceminatum* Wall., *Rhodendran Campamelatum* D. Don.	High levels	Rock fall, Avalanche

Table 2. Species composition and number of individuals of butterflies from different study sites in the Pindari are recorded during 2003-05.

Sl. No	*Taxonomic Composition*	*2003-2005*			*Total*
	LEPIDOPTERA	*Site - 1*	*Site-2*	*Site-3*	
	Nymphalidae				
1.	*Vanessa cashmirensis* Fru.	164	94	40	298
2.	*Vanessa indica indica* Herbest	10	50	42	102
3.	*Precis iphita siccata* (Stichel)	58	35	29	122
4.	*Pyramies cardui cardui* Linn.	12	40	26	78
5.	*Pseudoergoils wedah* Kollar	24	10	-	34
6.	*Neptis yerburyi yerburyi* (But.)	74	45	45	164
7.	*Argynnis lathonia issea* Doubleday	-	14	41	55
8.	*Atella phalanta phalanta* Drury	10	-	-	10
9.	*Sephisa dichroa* (Kollar)	55	18	-	73
10.	*Rahinda hardonia* Stoll.	12	3	-	15

	Pieridae				
11.	*Pieris canidia indica Sparr.*	95	82	55	232
12.	*Pieris brassicae* Linn.	60	60	27	147
13.	*Gonepteryx rhmni neplensis* Doubleday	65	46	25	136
14.	*Delias b. belladonna* Fabr.	20	22	11	53
15.	*Aporia agathon caphisa* Moore	75	44	35	151
16.	*Anapheis aurota aurota* (Fabr.)	13	—	—	13
17.	*Colias electo fieldi* Menestries	67	49	35	151
18.	*Colias erate* (Esper.)	—	19	25	46
19.	*Huphina herissa phryne Fabr.*	29	—	—	29
	Lycaenidae				
20.	*Heliphorous sena* (Kollar)	74	72	33	179
21	*Hedoes pavana* (Kollar)	58	53	47	158
22.	*Celastrina huegelli* Moore	42	10	6	58
23.	*Chilaria kina* (Hewiston)	11	—	—	11
24.	*Lampides bocticus* (Linn.)	33	13	2	48
25.	*Virachola issocrates* F.	20	18	5	43
	Satyridae				
26.	*Aulocera swaha swaha* (Kollar)	65	48	35	148
27.	*Pararge schakra* Koli.	50	33	13	96
28.	*Callerebia scanda scanda (Kollar)*	38	30	21	89
29.	*Lethe sidonis sidonis* Hew.	22	28	16	66
30.	*Ypthima* sp.	49	21	2	72
	Papilionidae				
31.	*Paridis alidoneus* Doubday	38	31	25	94
32.	*Papilio polytes romulus* Cramer.	20	—	—	20
33.	*Byasa varuna astorian* Wd.	30	32	6	68
	Danaidae				
34.	*Euploea core* (Cramer)	33	5	—	38
35.	*Danaus aglae* Stoll.	21	—	—	21
	Erycinidae				
36.	*Dodona durga* (Kollar)	12	43	33	88
	Acraeidae				
37.	*Acraea issoria anomal* Kollar	22	16	36	74
	Sphingidae				
38.	Macroglossum sp.	22	—	—	22
39.	*Panaera mydon* Walk.	—	—	11	11

	Lithosiidae				
40.	*Chrysorabdia buttata* Walk.	24	3	4	31
41.	*Agylla rufifrons* Moore	7	—	—	7
	Noctuidae				
42.	*Calpe ophideroides* Guen.	19	26	—	45
	Zygaeinidae				
43.	*Chaleosia pectinicomis* albata Moore	—	23	20	43
	Notodontidae				
44.	*Gazalina chrysolopha* Koll.	27	15	—	42
	Syntomidae				
45.	*Ceryx imaon* Cramer	34	19	18	71
	Arctiidae				
46.	*Estigmene* sp.	11	5	—	16

Table 3. Community structure of butterflies across sites during the study period.

Ecological Units	*Diversity (Shannon-Wiener)*	*Evenness (Shannon-Wiener)*	*Richness (Total number of species)*	*Abundance (Total number of individuals)*
Site 1	3.325	0.751	42	1625
Site 2	3.146	0.738	37	1175
Site 3	2.838	0.704	31	892

Species Composition of Butterflies

Species composition and number of individuals of insects collected from different study sites has been given in Table 2. A total of 46 species of butterflies (Lepidoptera) belonging to 15 families were recorded from the study sites. Family Nymphalidae with 10 species and 1091 individuals was the most dominant family. Nymphalidae constituted 27.73% of total collected Lepidoptera. Out of 10 species of this family *Vanessa cashmirensis* Fru was the most dominant species which constituted 31.65% of total individuals of this family followed by *Neptis yerburyi yerburyi* (But.) (16.58%), *Precis iphata siccata* Stichel (12.34%), *Vanessa indica indica* Herbest (10.11 %) and *Sephisa dichora* (Kollar) (9.91 %).Pieridae with 09 species and 958 individuals was the second largest family constituting 26.53% of the total. *Pieris canida indica* Sparrman was the most dominant species of this family constituting 23.15% of total individuals of this family, followed by *Aporia agathon caphisa* Moore (15.96%), *Calias electo fieldi* Menestris (15.96%) and *Pieris brassicae* Linn. (15.54%).

Family Lycaenidae with 06 species and 497 individuals constituted 13.77% of the total collected Lepidoptera. *Heliphorous sena* (Kollar) with (36.46%) was the most dominant species of this order followed by *Heodoes pavana* (Kollar) (32.18%) and *Celastrina huegeilli* Moore (10.59%). Family Satyridae constituted 12.59% of total and *Aulocera swaha swaha* (Kollar) was the dominant species of this family. Family Papilionidae was represented by three species and constituted 5.10%. *Paridis alidoneus* Doubleday was the dominant insect of this family. *Vanessa cashmirensis* Fru. (Nymphalidae) was the most abundant species present in all the three study areas constituting 8.1 % towards total number of individuals, collected from all the three areas. It was followed by *Pieris canidia indica* Sparr. (Pieridae) (6.3%) and *Heliphorous sena* (Lycaenidae) *(4.8)*. Eight species of butterflies were confined to site at lowest altitude only these included *Atella phalanta phalanta* Drury, *Anaphaeis aurota aurota* (Fabr), *Huphina herissa phryne* (Fabr.), *Chi/aria kina* (Hewiston), *Papilio polytes romulus* Cramer., *Danaus aglae* Stoll., *Macroglossum* sp., *Agylla rufifrons* Moore. Six species of lepidopteran insects were confined to sites 1 and 2 which included *Pseudoergolis wedah* Kollar, *Sephisa dichroa* Kollar, *Rahinda hardonia* Stoll, *Euploea core* (Cramer), *Calpe ophideroides* Guen , *Gazalina chrysolopha* Koll. hid *Estigmenesp.* While three species were recorded from area 2 and 3 only viz. *Argynmis lathonia issea* Doubleday, *Calias erate* (Esper.) and *Chaleosia pectinicornis albata* Moore. The rare species included *Altla phalanta phalanta* Drury, *Chi/aria kina* (Hewiston) and *Panaera mydon* Walk. *Panaera mydon* Walk was confined to Phurkia area only.

Site 1 which was at the lowest altitude supported as many as 42 species and 1625 individuals representing 91.3% and 44.0% of total number of species and individuals recorded from the Pindari area. In this study area *Vanessa cashmirensis* Fru. was the most dominant species constituting 10.1 % of total individuals. It was followed by *Pieris canidia indica* Sparr. (5.84%), *Heliphorous sena* Kollar (4.5%) *Gonepteryx rhamni neplensis* Doubleday and *Aulocera swaha swaha* (Kollar) (4% each). In this study area *Agylla rufifrons* Moore with 7 individuals was the most rare species constituting 0.43% of total. This species was recorded from this area only. Other species which were collected on few occasions included Vanessa *indica indica* Herbest, *Atella phalanta phalanta* Drury and *Chilaria kina* (Hewiston). Similarly, site two supported 37 species and 1175 individuals which represented 80% and 31.8% of total species and individuals recorded from the area respectively. *Vanessa cashmirensis* Fru. was again the most dominant species constituting 8% of total individuals collected from this area. It was followed by *Pieris canidia indica* Sparr (7%) and *Heliphorous sena* Kollar (6.3%). Other dominant species of this area included *Heodoes pavana* (Kollar) and *Vanessa indica indica* Herbest. The least abundant species in this area were *Rahinda hardonia* Stoll., *Chrysorabdia buttata walk and Euploea core* (Cramer) and *Estigmene* sp.

Site three, at the highest altitude was represented by 31 species and 892 individuals which were 67% and 31 % of total species and individuals. *Pieris canidia indica* Sparr was the most dominant species in this area with 55 individuals constituting 6.2 % of total individuals collected from this area. It was closely followed by *Heodoes* pavana (Kollar) (5.3%) , Vanessa *indica indica* Herbest (4.7%) and Vanessa *cashmirensis* Fru. (4.48%). The least abundant species included *Lampides bocticus* (Linn.), *Ypthima* sp., *Chrysorabdia buttata* Walk and Byasa *varuna astorian* Wd.

Community Structure

Data on community structure of butterflies have been presented in Table 3. The ecological indices show a clear difference among the habitats. Diversity, a function of plant community, climatic conditions and disturbances, was highest (3.325) for Site 1. Similarly the evenness also tracked the diversity and highest value was recorded for site 1. The impact of severe climatic conditions was quite evident on the community structure of butterflies as the lowest value of all the parameters were obtained for site 3.

Discussion

Plants, animals and microorganisms are mostly considered to be separate entities, thereby largely ignoring the fact that most organisms do not live on their own, but depend on more or less intimate interactions with other species[3]. Similarly, being highly mobile, chances of their getting distributed over a wide geographical region are high. The forest habitats studied in the present investigation had different combinations of plant communities and climatic conditions associated with the very low to very high level of disturbances, hence a marked difference in the species composition, abundance, diversity and evenness is recorded among the different sites. In the present study the highest number of species have been recorded from the site 1, which is a moderately disturbed site at the lowest altitude. This may be due to the role of environmental factors such as temperature, humidity and, heterogeneity of the habitat, which determine the survival of the species for a longer period. Effect of minimum plant diversity and high disturbances have affected the diversity, evenness, and abundance of butterflies in site 3 as is evident from their community structure in this site. Several studies have shown that disturbance is an important mechanism, maintaining the species diversity[4-5]. Higher values of diversity and species richness of butterfly community have been reported in rural communities near a village and in large clearings on the forest edge than in the closed forest near the Tam Dao Mountains in Vietnam[6].

Southwood[7] has reviewed the key factor studies of insect, and suggested that the most wood land natural habitat herbivores, specially the sap suckers are more influenced by the food quality. Similarly, it has been found that 75% of capture of neotropical butterflies were from open areas as compared to 25% capture from the closed microhabitats[8]. These sampling differences reflect a community wide pattern of increasing insect activity with increasing light and temperature. This view is further supported by the presence of maximum number of species and highest abundance of butterflies during summer season in the present investigation.

As in case of other organisms, changes in the climatic or edaphic conditions are known to affect the butterflies. Many butterflies can remain dormant as larvae or pupae over long periods especially during adverse weather conditions and their seasonal forms occur in the habitats characterized by changing environmental conditions.

In a survey from the old Kumaun region 371 species of butterflies have been recorded, which now includs the present Garhwal region with 8 districts and the Kumaun region with 5 districts[9]. As many as 835 species of butterflies have earlier been recorded from eastern Himalayas and 415 species from Western Himalayas[10]. The effect of vegetation, climatic factors such as temperature and rainfall on Lepidoptera has also been studied by different workers[11-14].

The studied habitats fall in a very sensitive eco-climatic region of Western Himalaya. The results obtained here indicate that the small habitats surveyed in the present investigation have a high butterfly diversity and richness and regular surveys and monitoring of such sensitive area is essential for the conservation of ecologically important species.

References

1. Southwood, T.R.E. 1978. In: *Ecological Methods.* Chapman and Hall, New York.
2. Kotila, P.M. 1986. *Ecological Measures.* Software package, St. Lawrence University, USA.
3. Redfern, A. and Pimm, S.L. (1987). In: Insect outbreaks and community structure. In: Insect outbreaks (eds. P. Barbosa & J.C.Schultz), p. 99. Academic Press, San Diego.
4. Jacobs, M. (1988). The tropical rainforest. A first encounter (Berlin: Springer-Verlag).
5. Huston, M.A. (1994). In: *Biological diversity.* The coexistence of species on changing landscapes (Cambridge University Press).
6. Leps, J. and Spitzer, K. (1990). *Acta Entomol.* 87: 182.
7. Southwood, T.R.E. 1975. The dynamics of incest populations. pp. 151-199. In: D Pimental (ed.) *Insects, Science and Society.* Academic Press, New York.
8. Srygley R.B. & P. Chi. 1990. *The American Naturalist.* 135: 765-787.
9. Hannygton, F. (1910). *Journal of Bomb. Nat. Hist. Soc.* 20(1): 130.
10. Wynter-Blyth, M.A. (1957). *Butterflies of the Indian region:* XX + pp. 523 (Published by BNHS, Bombay).
11. Pollard, E. (1988). *J. Appl. Ecol. 25: 819-828.*
12. Bergelson, J.M. and Lawton, J.H. (1988). *Ecology*, 62 (2).: 434-445.
13. Mani, M.S. (1956). *J. Bom. Nat. Hist. Soc.* 58(3)724-748.
14. Beeson, C.F.C. (1941). In: The Ecology and Control of the forest insects of India and the neighbouring countries. Vasant Press, Dehradun. pp. 1-1007.

5
Management of Jassid *Empoasca spinosa* Dworowaska and Sohi in Fenugreek Through new Seed Dressers

M.N. KAPADIA AND P.G. BUTANI
Department of Entomology, College of Agriculture, Junagadh Agricultural University, Junagadh - 362 001

Abstract

A field experiment was conducted to evaluate the dose response of two new molecules (imidacloprid 70 WS and thiamethoxam 70 WS) as seed dresser in comparison to the conventional insecticides, dimethoate @ 0.03 per cent (foliar application) against the jassid (*Empoasca spinosa* Dworowaska and Sohi) of fenugreek at the Junagadh Agricultural University Campus, Junagadh during two winter seasons, 2004-05 and 2005-06. The results showed that the higher dose of imidacloprid @ 10 g/kg seed and thiamethoxam @ 4.2 g/kg seed were found effective up to 90 days after germination. The seed treatment of thiamethoxam @ 4.2 g/kg seed gave significantly highest seed yield (1827 kg/ha) and net profit (Rs. 3300), followed by imidacloprid @ 10 g/kg seed (1798 kg/ha) and Rs. 2952. However, incremental cost benefit ratio was obtained highest (ICBR, 1:2.12) in thiamethoxam @ 2.8 g/kg seed, followed by thiamethoxam @ 10 g/kg seed (1:1.98) and imidacloprid @ 10 g/kg seed (1:0.70). Thus thiamethoxam @ 4.2 g/kg seed and imidacloprid @ 10 g/kg seed were found the most effective and economical seed dressers for the management of jassids in fenugreek.

Introduction

Fenugreek, important seed spices crop is attacked by sucking pests mainly jassid (*Empoasca spinosa* Dworowaska and Sohi). It causes damage more or less through out the crop season results in to quantity and quality losses. A perusal literature has shown that no work has been done on the insecticidal seed treatments for its control in fenugreek. Imidacloprid and thiamethoxam as seed dressers have previously been recommended for initial sucking pests of okra and cotton. To minimize the insecticides load in fenugreek ecosystem, seed dressing of fenugreek seed with new molecules is an immediate need as an alternative of hazardous foliar application of insecticides. Present study was conducted to evaluate the new molecules as seed dressers against jassid in fenugreek.

Materials and Methods

The fenugreek seeds were treated with different doses of imidacloprid 70 WS and thiamethoxam 70 WS. Both insecticides were diluted in water (1:1 ratio) and starch solution was added. Seeds were treated in polythene bag and stirred well for uniform coating. The experiment was conducted in a randomized block design replicated thrice with plot size of 5.0m X 2.7m during two winter seasons, 2004-06. The conventional insecticide, dimethoate 0.03% was compared with seed treatments. Five plants selected randomly from each plots were observed to record the jassid population on three leaves (upper, middle and lower) from each plant at 10 days interval up to 90 days after germination.

Results and Discussion

Data (Table 1) showed that significantly lowest jassid population was recorded in thiamethoxam @4.2 g/kg seed which was at par with imidacloprid @ 10 g/kg seed, followed by thiamethoxam @2.8 g/kg seed. Both the seed dressers remained comparatively most effective as compared to the foliar application of dimethoate 0.03 per cent. The result based on overall mean pest population indicated that both the seed dressers at higher dose were found effective to reduce the pest upto 90 days after germination. The thiamethoxam @ 4.2 g/kg seed recorded higher seed yield (1827 kg/ha) which was at par with imidacloprid @ 10 g/kg seed (1798 kg/ha) and thiamethoxam 2.8 g/kg seed (1748 kg/ha). Highest net realization (Rs. 3300/ha) was obtained in thiamethoxam @ 10 g/kg seed, followed by imidacloprid @ 10 g/kg seed (Rs. 2952/ha). Whereas, highest ICR was obtained in thiamethoxam @ 2.8 g/kg seed, followed by thiamethoxam @ 4.2 g/kg seed (1:1.98) and imidacloprid @ 10 g/kg seed (1:0.70). Imidacloprid @ 10 g and thiamethoxam 4.2 g/kg seed has been reported most effective seed dressers against sucking pests of cotton[1-3]. Mittal and Butani[4] previously recommended the insecticide sprays at 30 days interval starting from 30 days after germination of fenugreek for effective management of *Empoasca spinosa*.

The study concluded that thiamethoxam 4.2 g/kg seed and imidacloprid 10 g/kg seed were found the most effective and economical for management of jassid in fenugreek.

Table 1. Effect of insecticidal seed treatments on jassid population and yield of fenugreek and their economics (pooled of 2004-05 and 2005-06)

Treatment	*Jassids / 3 leaves / Plant* 60 DAG	70 DAG	80 DAG	90 DAG	*Mean*	*Grain yield, kg /ha*	*Net realization Rs/ha*	*ICBR*
1. Imidacloprid 70WS @ 5g/kg seed	1.00 (0.49)	1.43 (1.56)	1.65 (2.22)	1.69 (2.35)	1.44 (1.58)	1610	696	1:0.33
2. Imidaclorpid 70WS @ 7.5g/kg seed	0.93 (0.36)	1.31 (1.22)	1.42 (1.52)	1.50 (1.76)	1.29 (1.17)	1687	1620	1:0.51
3. Imidacloprid 70WS @ 1 0g/kg seed	0.85 (0.23)	1.08 (0.66)	1.06 (0.62)	1.15 (0.83)	1.03 (0.57)	1798	2952	1:0.70
4. Thiamethoxam 70 WS @ 2.8g kg seed	0.81 (0.15)	1.14 (0.79)	1.22 (0.99)	1.33 (1.26)	1.12 (0.76)	1748	2352	1:2.12
5. Thiamethoxam 70WS @ 4.2g kg seed	0.81 (0.16)	0.96 (0.43)	0.87 (0.26)	1.09 (0.69)	0.94 (0.37)	1827	3300	1:1.98
6. Dimethoate 30 EC 0. 03 % on	1.08 (0.66)	1.78 (2.68)	2.28 (4.69)	2.51 (5.78)	1.91 (3.15)	1570	216	1:0.42
7. Control (water spray)	1.15 (0.82)	1.76 (2.59)	2.28 (4.72)	2.45 (5.52)	1.91 (3.15)	1570	216	–
8. Control (Untreated)	1.12 (0.76)	1.82 (2.82)	2.28 (4.69)	2.53 (5.89)	1.94 (3.25)	1552	–	–
S.Em.±	0.04	0.05	0.05	0.05	0.12	45.73		
C.D. at 5%	0.11	0.14	0.14	0.16	0.34	132.5		
C.V. %	9.47	8.60	7.46	7.45	15.88	6.71		

$\sqrt{x + 0.5}$ transformation used. Data in parentheses are retransformed values. DAG= Days after germination.

References

1. Vadodaria, M.P.; Patel C.J.; Maisuria, I.M. and Patel, U.G. (2001). *G.A.U. Res.* J., 26 (2): 32.
2. Bheemanna, M.; Patil, B.V.; Anand Kumar, V. and Hanchinal, S.G. (2003). In:Bio efficacy of thiamethoxam seed treatments against cotton early sucking insect pests. Proced. National Symposium on Frontier Areas of Entomological Research, 5-7 November at New Delhi, p. 169.
3. Kapadia, M.N., and Butani, P.G. (2005). Eco friendly management of sucking pests in summer okra through new seed dressers. Proced. National Seminar on Applied Entomology: Current Status, Challenges and Opportunities, 26-28 September, Udaipur, p. 109.
4. Mittai, V.P. and Butani, P.G. (1989). Bio efficacy of certain insecticides against jassids attacking fenugreek. Paper abstract in 1st National Symposium on Seed spices, 24-25 October, Udaipur, p. 39.

6

Population Dynamics and Seasonal Abundance of Rotifers in A Shivalik Stream (Ban Ganga)

M.K. JYOTI, K.K. SHARMA, NITASHA SAWHNEY AND SARBJEET KOUR

Department of Zoology, University of Jammu, Jammu-180006

Abstract

The present communication deals with the seasonal abundance and distribution of rotifer fauna in a shallow snow fed Ban Ganga stream (Katra). The twelve species of rotifers recorded during present investigations (Jan., 2003 - Dec., 2003) belong to two classes Monogononta and Digononta with maximum representation of order Ploima (10 sps.) of class Monagononta. Various physico-chemical parameters were also analysed to assess their influence on the seasonal distributional trend among rotifer inhabiting the stream.

Introduction

Rotifera, a group of dominant plankters among zooplankton community, constitutes an important interlace in the aquatic food chain. Though these are microcreatures yet their role as food in early larval stages of fish where these are known to save the fish biopotential and use as indicators for assessing water quality is immense. Thus, keeping in view, their ecological* importance, the study of these organisms becomes highly important from point of view of their diversity and seasonal dynamics. Leeuwenhoek[1] was the first to have made observation about these organisms. Today, these organisms are regarded as cosmopolitan in distribution and various aspects of their biology and life cycle have been studied[2-8]. In the state of Jammu and Kashmir some information on the diversity of rotifers has been raised in this part[9-18] and a lot remains to be done. Recognizing the need, therefore an attempt has made to record their presence and distribution in snow fed Ban Ganga stream.

Materials and Methods

Water samples were collected on monthly basis from Jan., 2003-Dec., 2003 from stations I-III. These stations were raised along the course of stream, station II is 0.2 km downstream from Station-I and Station-III 1.5 km. downstream station II.

Each sample was analysed for some physico-chemical parameters (APHA)[19]. Collection of zooplankton for the study of rotifers was done by filtering 50 litres of water through plankton net (bolting silk no. 25, mesh size 0.03 μm). The collected samples were preserved using 5% formalin and studied in the laboratory. Laboratory procedures involved sorting and identification using Olympus microscope. Rotifer species were counted using Sedgwick-Rafter Cell and their number was calculated as

$$\text{Number/l} = \frac{C \text{ x } 1000m^3}{A \times D \times F}$$

where, C = Number of organisms counted.

A = Area of a field.

D = Depth of field (S-R cell depth-1m)

F = Number of fields counted.

Results and Discussion

Abiotic Characteristics

In order to analyse the influence of abiotic factors on the seasonal distributional trend among rotifers inhabiting the stream, various physico-chemical parameters were assessed. Perusal of Table 1 reveals that a wide range of fluctuation appeared in these parameters. A close relationship between water and air temperature was recorded. pH of the stream recorded fluctuations ranging from minimum of 8.1 to a maximum of 8.8. Dissolved oxygen fluctuated from 6.4 mg/l (January) to 8.4 mg/l (August) and FCO_2 remained absent throughout the study period while the carbonates fluctuated from 23.0 mg/l (October) to 41.0 mg/l (April). The fluctuations in bicarbonate was from 57.9mg/l (November) to 242.9 mg/l (March).

Similarly, chloride fluctuated from 15.2 mg/l (September) to 27.2 mg/l (May). Calcium ranged from 26.6 mg/l (July) to 41.3 mg/l (January) and magnesium from 20.2 mg/l (July) to 29.3 mg/l (April).

Rotifer Distribution

The twelve species of rotifers collected has been categorized into two classes, Monogononta and Digononta. Of the twelve species of rotifers, only one species *(Philodina* Sps.) belonging to family philodinidae of order Bdelloidea under class Digononta whereas all other (11 species) belong to six families (Notommatidae, Brachionidae, Colurellidae, Lecanidae,) Filinidae and Dicranophoridae of 2 orders Ploima (10 species) and Flosculariacea (1 species) under class Monogononta. Thus, depicting the maxima of order Ploima in lotic waters[20].

Rotifera inhabiting Ban Ganga stream:

Class: Monogononta	**Class : Digononta**
Order: Ploima	**Order: Bdelloidae**
Brachionidae	**Philonidinidae**
Brachionus bidentata	*Philodina* sps.
Notommatidae	
Cephalodella gibba	
Lecaniidae	
Lecane *bulla, L. ludwigi, L. closterocerca, L. luna*	
Colurellidae	
Colurella obtusa	
C. uncinata	
Lepadella ovalis	
Dicranophoridae	
Dicranophorus epicharis	
Order : Flosculariacea	
Filiniidae	
Filinia longiseta	

12 species of rotifers recorded presently exhibited a well marked seasonal and spatial distribution (Table 2, 3 & 4). A look at the Table 5 Fig. 1 reveals that of the three study stations investigated presently, station I contributed 23.8% of rotifers to total rotifer count, represented by 5

Table 1. Seasonal variation in physico-chemical parameters of Ban Ganga stream Jan., 2003- Dec., 2003

Months	*Air Temp. (°C) & weather*	*Water Temp. (°C)*	*Depth (cm)*	*Veloc-ity (m/s)*	*PH*	*DO (mg/l)*	*FCO_2 (mg/l)*	*CO_3 (mg/l)*	*HCO_3 (mg/l)*	*Cl (mg/l)*	*Ca (mg/l)*	*mg (mg/l)*
J	19.0 PC	16.0	20.0	0.04	8.1	6.4	-	27.0	223.8	22.5	41.3	29.0
F	15.0 S	16.3	19.3	0.04	8.3	7.4	-	33.0	221.0	17.9	37.3	28.9
M	24 .0S	19.3	19.0	0.04	8.8	7.7	-	34.0	242.9	20.5	36.5	26.4
A	26.0 S	17.8	21.3	0.04	8.7	7.4	-	41.0	185.9	21.9	33.0	29.3
M	32.1 S	27.3	18.0	0.04	8.1	7.6	-	35.5	187.0	27.2	37.6	28.4
J	30. 6 PC	28.0	18.6	0.03	8.5	7.5	-	35.0	209.4	23.9	32.0	22.6
J	30.3 S	27.0	23.6	0.05	8.6	8.0	-	29.0	162.6	21.2	26.6	20.2
A	28.8 PC	24.3	30.3	0.09	8.2	8.4	-	25.0	112.8	22.5	36.2	24.2
S	27. 6 PC	24.0	30.6	0.09	8.2	7.8	-	40.0	67.0	15.2	37.6	27.6
O	24.3 S	22.0	21.0	0.06	8.5	7.4	-	23.0	82.3	20.5	34.9	25.0
N	22.0S	19.0	18.3	0.04	8.5	7.6	-	23.0	57.9	15.9	32.0	22.1
D	18.0 S	16.8	18.6	0.04	8.1	7.0	-	25.0	65.0	15.9	38.4	22.6

• S= Sunny and PC = Partially cloudy

species *(Philodina sps., Lepadella ovalis, Colurella uncinata, Lecane bulla. Lecane closterocerca)*, station II contributed 49.7% rotifers to total rotifer count represented by 10 species *(Philodina sps., Lepadella avails, Colurella obtusa., C. uncinata, Lecane bulla, L. ludwigi, L. luna, L. closterocerca, Dicranophorus epicharis and Filinia longiseta)* and station III contributed 26.3% rotifers to the total rotifer count represented by 7 species *(Brachionus bidentata. Philodina sps. Lepadella ovalis, Colurella uncinata, Cephalodella gibba, Lecane bulla and L. closterocerca)*.

Perusal of Table 6 reveals that different species inhabiting each station exhibited different contributional percentage to the total rotifers of that station. Of the 5 species recorded at station I, *Lepadella ovalis* contributed maximally (28.8%) and *Colurella uncinata* minimally (9.3%) whereas among the 10 species of rotifers recorded at station II the maximum contribution was of *Philodina* species (50.6%) and minimum contribution was of *Lecane luna* (0.4%) and at station III dominance was again exhibited by *Philodina species* (29.4%) alongwith *Lepadella ovalis* (29.4%) and minimum percentage was shown by *Brachionus bidentata and Colurella uncinata* (0.7%).

Perusal of Table 7 reveals that rotifers also show quantitative fluctuations seasonally in Ban Ganga stream ranging from minima of 100 + 40.8n/l in February to maxima of 2450±1181.1 n/l in November. The early winter maxima in rotifer population may be due to influence of many factors such as dissolved oxygen (DO) concentration, absence of FCO_2, alkalinity and presence of other optimal conditions (biotic) for survival[20]. The late winter minima in rotifer population may be due to the changes in the abiotic components (Table 1) and biotic components of water body[21]. Differential quantitative and qualitative distribution of rotifers at 3 stations was observed during present course of investigation (Table 6), the sequence being station II> station III> station I. Thus emphasizing that although the rotifers are present in all types of water bodies, the diversity is best flourished in lentic conditions with less speed[20] as station II exhibit lentic conditions (slow water speed and increased depth) due to embankment. Furthermore, from the recorded data, it is emphasized that species like *Dicranophorus epicharis, Filinia longiseta, Lecane ludwigi* and *L. luna* were exclusively present at station II and *Brachionus bidentata* and *Cephalodella gibba* at station III.

Table 2. Seasonal abundance of Rotifers (n/e) of Ban Ganga stream at station I (Jan., 2003-Dec., 2003)

Organisms / Months	Jan.	Feb.	Mar.	April	May	June	July	Aug.	Sep.	Oct.	Nov.	Dec.	Total	Mean±SD
Rotifer A														
Brachionus bidentata														
Philodina sps.						200	550				50		800	66.6±155.9
Lepadella ovalis	300		50				100		100		1050	100	1700	141.6±286.3
Colurella obtusa														
C. uncinata											350	200	550	45.8±106.9
Filinialongiseta														
Dicranophorus epicharis														
Cephalodella gibba														
Lecanebulla	50	60	50		50						1000		1200	100.0 ± 272.3
L. closterocerca											1650		1650	137.5± 456.0
L. ludwigi														
L. luna														
Total Rotrfera	350	50	100		50	200	650		100		4100	300		

Table 3. Seasonal abundance of Rotifers (n/e) of Ban Ganga stream at station II (Jan., 2003-Dec., 2003)

Organisms / *Months*	*Jan.*	*Feb.*	*Mar.*	*April*	*May*	*June*	*July*	*Aug.*	*Sep.*	*Oct.*	*Nov.*	*Dec.*	*Total*	*Mean ±SD*
Rotifer A														
Brachlonus bidentata														
Philodina sps.		100	300	400	450	3750	500	100	50	250	300		6200	5166 ±9889
Lepadella ovalis	350					50	150				100	250	900	75.0 ± 112.7
Colurella obtusa										200			200	16.6 ± 55.2
C. uncinata										100		50	150	12.5 ± 29.7
Filinia longiseta								100					100	8.3 ± 27.6
Dicranophorus epicharis											100		100	8.3 ± 27.6
Cephalodella gibba														
Lecane bulla				150	100			100	50	2000	300		2700	225.0 ± 542.1
L.closterocerca							50				550	50	650	541.1±150.6
L. ludwigi						1200							1200	100 ± 331.6
L. luna												50	50	4.16 ±13.8
Total Rotifera	350	100	300	550	550	500	700	300	100	2550	1400	350		

Table 4. Seasonal abundance of Rotifers (n/l) of Ban Ganga stream at station III (Jan.. 2003-Dec., 2003).

Organisms / Months	Jan.	Feb.	Mar.	April	May	June	July	Aug.	Sep.	Oct.	Nov.	Dec.	Total	Mean±SD
Rotifer A														
Brachlonus bidentata	50												50	41 ± 138
Philodina sps.		100	50		400		200	300	200	300	350		1900	1583.3±145.5
Lepadella ovalis	400		100					100			1300		1900	158.3±361.6
Colurella obtusa														
C. uncinata											50		50	4.1±13.8
Filinia longiseta														
Dicranophorus epicharis														
Cephalodella gibba								300	300				600	50.0±118
Lecane bulla	100	50		50	50	1000		50	50		150		1500	125.0±267.3
L. closterocerca										450			450	37.5± 124.3
L. ludwigi														
L. luna														
Total Rotifera	500	150	150	50	450	1000	200	800	550	750	1850			

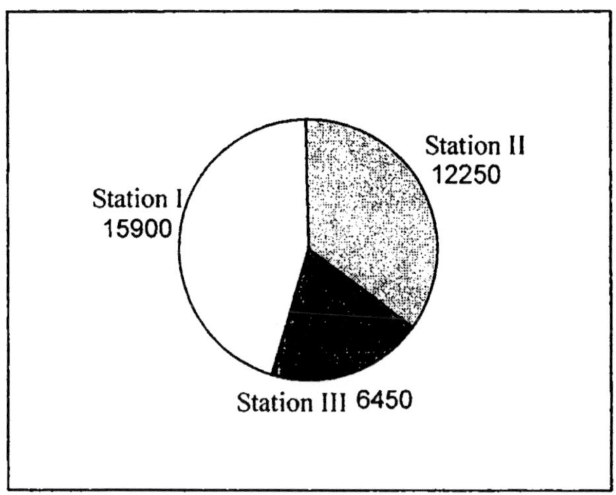

Fig. 1: Percent contribution of Rotifers to total Rotifer community along the three study stations of Ban Ganga stream.

Table 5. %Contribution of Rotifers to total Rotifer community.

Organisms	*Total number n/1*	*% contribution to total rotifer community*
Station I	5900	23.8%
Station II	12250	49.7%
Station III	6450	26.2%
	24600	

It is also inferred that *Lepadella ovalis* and *Lecane bulla* are the two species which can tolerate low levels of oxygen as these were recorded from all the stations where DO content was minimum[22]. Perusal of Table 3 also enlights the dominant nature of group Lecaneidae which showed the presence of four species *(Lecane bulla, L. clostemcerca, L. ludwigi* and L. *luna).* The observations regarding presence of three species of *Lecane (L. bulla, L. closterourca and L. luna)* in the month of November at S-II of stream is in concordance with the earlier findings[23-24].

Table 6. % contribution of individual sps. to the total Rotifer population

Organisms	Station I I	% contribution of ind. Sps.to the total Rotifer population	Station II	% contribution of ind. Sps.to the total Rotifer population population	Station III	% contribution of ind. Sps.to the total Rotifer population population
Rotifer A						
Brachionus bidentata					50	0.7%
Philodina sps.	800	13.5%	6200	50.60%	1900	29.4%
Lepadella ovalis	1700	288%	900	7.3%	1900	29.4%
Colurella obtusa			200	1.6%		
C. uncinata	550	9.3%	150	1.236	50	0.7%
Filinialongiseta			100	0.8%		
Dicranophorus epicharis			100	0.8%		
Cephalodella gibba					600	9.3%
Lecanebulla	1200	20.3%	2700	22.0%	1500	23.2%
L. closterocerca	1650	27.9%	650	5.3%	450	6.9%
L. ludwigi			1200	9.7%		
L. luna			50	0 4%		
Total	5900		12250		6450	

Table 7. Seasonal variations in the Rotifer population (n/l) along the three study stations of Ban Ganga stream (Jan., 2003 - Dec., 2003)

Months/Station	*Station I*	*Station II*	*Station III*	*Mean ± SD*
January	350	350	500	400 ± 70.7
February	50	100	150	100 ÷ 40. 8
March	100	300	150	1 83.3 ± 84.9
April	Ab	550	50	200 ± 284.3
May	50	550	450	350 ± 216.0
June	200	5000	1000	206.6 ± 2099.9
July	100	100	250	150 ± 70 7
August	Ab	300	800	366.6 ± 329.9
September	650	700	200	516.6 ± 224.8
October	Ab	2550	750	1100 ± 1070.0
November	4100	1400	1850	2450 ±11811
December	300	350	Ab	216.6 ± 154.5

References

1. Leeuwenhoek, A. (1703). In : *Animal Resources of India, Protozoa to Mammalia. Rotifera.*, 69. Edited by Director, Zoological Survey of India.
2. Hymen, L.H. (1951). In : *Invertebrates, Acanthocephala, Aschelminthes and Eutoprocata the pseudocoelomate bilateria,* Vol. III. McGraw Hill, New York, p. 572.
3. Edmondson, W.T. (1957). *Trans. Am. Microse. Soc.*, 76.
4. Lee, C.S., Hu, F. and Reijiro, H. (1981). *Prog. Fish. Cult.*, 43(3): 212.
5. Cruz, E.M. and James, C.M. (1989). *Aquaculture. 77* : 353.
6. Mookerjee, N. (1992). In : *Ph.D. Thesis, University of Delhi, India.*
7. *Ricci, C. (1992). In : Rotifera, Parthenogenesis and Heterogony, Sex, Origin and evaluation. R. Dllai* (ed.) Selected Symposia and Monagraphs U.Z. I, b, Mucchi, Modena, 329.
8. Claudia and Melone (2000). In : *Ph.D. Thesis* by Sharma Madhivi (2001).
9. Dutta, S.P.S. (1978). *Ph.D. thesis* Univ. of Jammu.
10. Zutshi, D.P. and Vass, K.K. (1978). *Ind. J. Ecol.*, 5 (1): 90.
11. Jyoti, M.K. and Sehgal, H.S. (1979). *a Hydrodiol., 65* : 23.

12. Gupta, V.K. And Sudan, A.V. (1985). *Bull. Env. Soc. 2* (2): 7.
13. Yousuf, A.R. and Mir, M.F. (1994). *J. Freshwater Biol.* 6 (2): 143.
14. Sharma, M. (2001). *Ph.D. Thesis,* University of Jammu, Jammu.
15. Baba, D.I. (2002). In : *Ph.D. Thesis,* University of Jammu, Jammu.
16. Sharma, S.P. (2002). In : *Ph.D. Thesis* submitted to the University of Jammu, Jammu.
17. Kour, S. (2002). In : *M. Phil. Dissertation* submitted to the Department of Zoology, University of Jammu.
18. Jan, N. (2005). In : *M.Phil, dissertation* submitted to Unversity of Jammu, Jammu.
19. A.P.H.A. (1985). In : *Standard methods of the examination of waste and waste water.* 16th ed. American Public Health Association, Washington, D.C.
20. Ricci, C. and Balsamo, M. (2000). *Reshw. Biol.,* 44 : 15.
21. Wetzel, R.G. (2001). In : *Limonology: Lake and River Ecosystems* (III edition). Academic Press. U.S.A. 1.
22. Prabhavathy, G. and Sreenivasan, A. (1977). In : *Proc. of sym. on warm water Zooplankton,* N.I.O., Goa, Spl. Publication: 319.
23. Jyoti, M.K. and Sehgal, H. (1980). *Limnologica.,* (Berlin). 1291.
24. Dalpatia, B.S. (1998). In : *Ph.D. Thesis,* University of Jammu.

7

Plankton in Relation to Fish Population in The River Suyal, Kumaun Himalaya

S.S. PATHANI, B.P. SINGH, A. KUMAR AND P.C. NAILWAL
Department of Zoology, Kumaun University Campus, Almora

Abstract

The spring fed river Suyal originates from Garuda Banz (2007 MSL) in the east of Almora, Uttarakhand State. It has several seasonal and perennial streams as its tributaries including Petshal and Sironiya streams. The river water harbours some 40 taxa of phytoplankton, 21 taxa of zooplankton and 19 species (5 families- Cyprinidae, Cobitidae, Channidae, Homalopteridae and Mastacembelidae) of fishes during the year, 2006. Monthly and seasonal rhythms of quality and quantity of phytoplankton and zooplankton with altitudinal variation are recorded for this river. The volume of phytoplankton and zooplankton on monthly basis ranged from 0.20 to 1.90 ml/l and 0.10 to 0.80 ml /l during the study. The annual average of volume of phytoplankton was 0.62 ml /l while the annual average of zooplankton was only 0.26 ml /l in the water. The ratio between the volume of phytoplankton and zooplankton is therefore, 2.31.

Introduction

The diversity of plankton and fish exert significant role in limnology of any freshwater ecosystem. The plankton not only supply food to the fry, fingerlings, young and adult fishes but also influence the abiotic features in the water[1-3]. Qualitative and quantitative estimation of plankton with abiotic features in Mountain Rivers has been, therefore studied[4,5]. Similarly, Tewari *et al.*[6] and Sharma and Sawhney[7] have collected a variety of phytoplankton and zoobenthos in the plains water of India. The present study encompass the qualitative and quantitative estimation of plankton, ratio of phytoplankton / zooplankton, gut analysis of some fishes (phytoplankton / zooplankton-ratio) in relation to the population of fishes in the water.

Materials and Methods

The samples of plankton were collected monthly in the river Suyal and its tributaries (Petsal garh, Sironiyagarh) by the plankton net for one year, 2006. The samples were brought to the Zoological laboratories of the Department and they were preserved in 5% formaline for volume and other studies [8,9]. The ratio between phytoplankton and zooplankton volume was conducted by centrifuging the plankton in separate centrifuge tube for each reading (Phytoplankton and zooplankton Ratio = P / Z: P = Phytoplankton volume in ml and Z = Zooplankton volume in ml). The gut contents analysis of fish was done by the volumetric method - Sedgewick Rafter, in terms of phytoplankton and zooplankton volume present in the intestinal bulb or in the stomach. The percentage of fish population was done on the basis of fish caught during the year in the study.

Results and Discussion

The river Suyal originates from Garuda Banj (2007 msl) The tributaries such as Petsal garh (1360 msl, 29° 37' 46" N and 79° 43' 38.39" E - Station I) and Sinoniya garh (1290 msl 29° 36' 7.01" N and 79° 43' 24.67" E - Station II) and many seasonal gadheras are leading into the river at Pithauni (1260 msl - Station III) and downwards.

Qualitative and quantitative estimation of plankton- The present study shows that there are 40 taxa of phytoplankton comprised of 15 genus of Chlorophyceae, 14 genus of Basillariophyceae , 9 genera of Cynophyceae and 2 genera of others (Table 1). The percentage of chlorophyceae ranged from 27.8 (August) to 59.3 (September), the bacillariophyceae ranged from 21.0% (June) to 41.0% (October) and the myxophyceae ranged from 12.2% (May) to 38.7% (September) in the water. There were 21 genera of zooplankton comparised of 11 genera of Protozoa, 5 genera of Rotifera and 5 genera of Arthropoda in the water (Table 2).

Table 1. Qualitative analysis of phytoplankton at different collecting spots in the River Suyal and its tributaries for one year

Phytoplankton	*Year- 2006*								
	Winter			*Summer*			*Monsoon*		
	I	*II*	*III*	*I*	*II*	*III*	*I*	*II*	*III*
Chlorophyceae									
Spirogyra	+	+	+	+	+	+	-	+	+
Ulothrix	+	+	+	-	+	-	-	+	-
Closterium	+	+	+	+	+	+	-	+	
Desmidium	-	+	+	+	-	+	+	+	
Cladophora	-	-	+	+	+	+	-	+	+
Zygnema	+	+	+	+	+	+	+	+	+
Microspora	+	-	-	+	+	+	-	+	+
Chlorococcus	-	-	+	+	+	+	+	-	+
Chlorella	+	+	+	-	-	+	+	-	+
Tetraspora	-	-	+	+	+	+	-	+	-
Ceratium	+	+	+	-	-	-	+	-	-
Schizogonium	+	+	+	+	+	+	+	+	+
Protococcus	-	+	-	-	+	+	-	-	+
Hormidium	+	-	+	+	+	+	-	-	-
Spirolaenia	+	+	+	+	+	+	-	+	-
Bacillariophyceae									
Cymbella	+	-	+	+	-	-	-	+	-
Navicula	-	+	-	-	+	+	-	+	+
Diatoma	+	+	+	+	+	+	+	+	+
Nitzchia	+	+	+	+	+	+	-	-	-
Neidium	-	-	+	+	+	+	+	+	+
Fragillaria	+	+	+	+	+	+	-	-	-
Metocera	-	-	+	+	+	+	+	-	-
Denticula	-	+	-	+	+	+	-	-	+
Ceratoneis	-	+	+	+	+	+	+	-	+
Pinnularia	-	+	+	+	+	+	-	+	+
Gomphoneis	+	+	+	+	+	+	-	+	+
Amphora	+	+	+	+	-	-	-	-	-
Cyclotella	+	-	-	+	+	+	-	+	+
Synedra	+	+	+	-	+	+	-	-	-

Contd.

Cynophyceae									
Anabaena	-	-	-	-	+	+	+	+	+
Rivularia	+	+	+	+	+	+	-	+	+
Oscillatoria	-	-	-	-	+	+	+	+	+
Microcystis	-	-	+	-	+	+	+	+	+
Lyngbya	-	-	+	-	-	-	+	-	-
Nostoc	+	-	-	-	-	-	+	-	-
Scytonema	-	+	-	-	-	-	-	+	-
Tolypothrix	+	+	-	-	-	+	-	-	-
Gloetrichia	-	-	+	-	-	-	+	+	-
Others									
Tribonema	+	+	+	-	-	+	+	+	+
Vaucheria	+	-	+	+	-	-	-	+	+

+ = Present and - = absent

The percentage of protozoans ranged from 2.3 (June) to 30.3 (September), the percentage of rotifers ranged from 6.5 (June) to 34.4 (November) and the percentage of arthropods ranged from 31.8 (November) to 82.3 (April) in per liter of the water. The availability of plankton varied monthly, seasonaly and station wise in the water. The population and site wise variations of plankton have been obtained in the present investigation (Table 1 and 2). The qualitative, quantitative, monthly, seasonally and altitudinal variations of plankton have also been recorded by various workers[10,13] in the lotic water of the Himalaya.

Plankton Volume: It is evident that the plankton total volume is always low in comparison to plains lotic waters which has shown monthly and station wise variations in the present study (Table 3 and 4).

The phytoplankton volume is always higher in summer months at all the stations than the zooplankton in the river and its tributaries. It is also recorded that the volume of phytoplankton is annually higher than the annual average volume of zooplankton (phytoplankton-0.62 ml/L and zooplankton- 0.26 ml/L). The phytoplankton volume Was high in summer season and it was moderate in winter but low during rainy season with bimodal rhythms followed by zooplankton in the water.

Table 2. Qualitative analysis of zooplankton at different collecting spots in the River Suyal and its tributaries for one year.

Zooplankton	*Year- 2006*								
	Winter			*Summer*			*Monsoon*		
	I	*II*	*III*	*I*	*II*	*III*	*I*	*II*	*III*
Protozoa									
Arcella	+	+	+	+	+	+		-	-
Euglena	-	-	+	+	+	- +		+	+
Cenfropyxis	+	-	+	-	-	-	+	+	+
Diffugia	-	+	+	+	+	+	+	+	+
Protodon	+	+	+	-	+	+	-	+	-
Bursaris	+	+	+	+	+	+	+	+	+
Vorticella	+	+	+	+	+	-	+	+	+
Volvox	+	+	+	-	+	+	+	+	+
Epistilis	+	+	-	+	-	+	-	-	+
Loxophyllum	+	-	-	-	+	+	+	+	+
Blepharisma	-	+	+	-	+	+	+	+	+
Rotifera									
Asplachna	+	+	+	-	+	+	+	+	+
Branchionous	+	+	+	+	+	+	+	+	+
Polyanthra	-	+	+	-	+	+	+	+	+
Keratella	+	-	+	+	+	+	-	+	+
Monostyla	+	+	+	-	-	+	+	+	+
Arthropoda									
Bosmina	+	+	+	+	+	+	-	-	+
Cyclops	+	+	+	-	-	+	+	-	+
Nauplius	+	+	+	+	+	+	+	+	+
Ceridophnia	+	-	+	+	-	+	-	-	+
Daphnia	-	-	+	-	-	+	+	+	-

+ = Present and - = absent

Table 3. Monthly volume (ml/l) of phytoplankton in the river Suyal and its tributaries during 2006.

	Jan	*Feb*	*Mar*	*Apr*	*May*	*Jun*	*Jul*	*Aug*	*Sep*	*Oct*	*Nov*	*Dec*	*Annual Average*
Spot 1	0.25	0.30	0.40	0.30	0.50	1.80	1.50	0.25	0.35	0.38	0.39	0.38	0.56
Spot 2	0.30	0.40	0.50	0.25	0.50	2.10	1.00	0.22	0.55	1.50	0.32	0.35	0.67
Spot 3	0.35	0.30	0.20	0.31	0.25	2.00	1.50	0.30	1.25	0.45	0.35	0.28	0.62

Table 4. Monthly volume (ml/l) of zooplankton in the river Suyal and its tributaries during 2006

	Jan	*Feb*	*Mar*	*Apr*	*May*	*Jun*	*Jul*	*Aug*	*Sep*	*Oct*	*Nov*	*Dec*	*Annual Average*
Spot 1	0.25	0.27	0.22	0.22	0.15	0.22	-	-	0.35	0.4	0.45	0.5	0.25
Spot 2	0.15	0.2	0.22	0.15	0.2	0.3	0.1	0.1	0.38	0.42	0.48	0.45	0.26
Spot 3	0.25	0.28	0.2	0.18	0.25	0.35	0.1	-	0.38	0.45	0.5	0.52	0.28

Annual ratio between phytoplankton and zoo plankton (P/Z) = 0.62 / 0.26 =2.31

Table 5. Showing fish population in % and P/Z ratio, in the gut contents

Fish Species	*Food Items ratio - P/Z*	*% of fish Population in the catch*
Barilius vagra	1.4	47.5
Garra lamta	1.8	26.2
Barbus chillinoides	1.0	6.6
Tor putitora	0.9	5.7
Mastacembelus armatus	0.7	4.9
Puntius sarana	0.9	4.0
Botia almorhae	0.5	2.0
others fishes	NA	4.0

The low volume of plankton may be due to high velocity of water as also reported in other rivers of the Himalaya[10,14]. A relationship between the annual volume of phytoplankton and zooplankton has been established as-P/Z = 2.31. It has a correlation to the gut contents of some fishes with their availability in the river and its tributaries. Fish fauna, fish population and gut analysis- Some 19 fish fauna have been identified during the study. They are included in five families *viz.* Cyprinidae-Tor tor, *T.putitora,*

Barbus chilinoides, Schizothorax richardsoni, S. plagiostomus, S. sinuatus, Garra gotyla, G. lamta, Puntius sarana, Barilius bendelisis, B. bendelisis chedra, B.vagra Cobitidae- *Nemacheilus botia, N. beavani, N. rupicola, Botia almorhae* Mastacembelidae- *Mastacembalus pancalus, M. armatus* Homalopteridae- *Homaloptera brucei* Channidae- *Channa gachua.* The population of these fishes varies in different region of the river Suyal and its tributaries. They migrate for feeding and breeding in different region of the water from downward to upward and vice versa in the deep and shallow areas in the lotic water bodies. Among 19 species of freshwater fishes identified, 7 species of fishes were analysed for their gut contents analysis in the study. The gut contents analysis on the basis of available phytoplankton and zooplankton of these 7 fishes have shown their population in the water for one year (Table 5). The highest population (47.5%) of Barilias *vagra* (herbivorous) and the lowest population (2.0 %) of Botia *almorhae* (carnivorous- Joshi, 2005) have been available in the water. The availability of other fishes are very few in the river|garhs and not sufficient to analyze the gut contents during the study which needs-further investigation.

There is a relationship between P/Z and fish population on the basis of P/Z values in the gut (Table 5). The phytoplankton availability is higher than zooplankton in the lotic water.The food and feeding habits of *Barbus chilinoides,Tor putitora* and *Puntius sarana* indicated omnivorous in feeding habit while *Mastacembelus armatus* and *Botia almorhae* are carnivorous or zoophagus in feeding habits, are low in population in the river. Therefore, it is presumed that the fish which feeds on phytoplankton is higher in number than the fish feeds on zooplankton in the river. Thus, the available food has a direct relation to the feeding habits of fish and growth population in the natural water body. Besides, the other fish present in the water are not sufficient to evaluate their feeding habits but this study indicates that the other fishes may be more zoophagus than the phytophagus.

References

1. Jhingran, V.G. (1991). In: Fish and fisheries of India, Hindustan Pub. Corporation (India) Dehli. p. 727.
2. Jhingran, V.G. and Sehgal K.L. (1978). In: Cold water fisheries of India. Inland fisheries society of India Barrackpore. p. 239.
3. Singh, H.R. (1993). In: Advances in Limnology. Narendra Publishing House, Delhi p. 366.

4. Pathani, S.S. Mahar, S. (2006). Flora and Fauna., 11(2): 250.
5. Pathani, S.S. Upadhyay K.K. (2006). In: An inventory on zooplankton, zoobenthos and fish fauna in the river Ramganga (W) of Uttaranchal, India. *Envis (*In press).
6. Tewari, V. Tripathi, S.K. and Chandel, B.S. (2006). Systematic enumeration of fresh water Chlorococcales of river Ganga at Kanpur Nagar (U.P.). Intn. Symp. on Current issues in Zoology and Environmental Science, Gorakhpur. p. 105.
7. Sharma, K.K. and Sawhney, N. (2006). In: Diversity of macrobenthic invertebrate fauna diversity in river Tawi. Intn. Symp. on Current issues in Zoology and Environmental Science, Gorakhpur. p. 57.
8. Ward, H.B. and Whipple, G.C. (1959). In: *Fresh Water Biology* 2nd Ed. John Wiley and Sons. New York, p. 1248.
9. Battish, S.K. (1992). In: Fresh Water Zooplankton of India. Oxford and IBH Publishing Co. Pvt. Ltd. New Delhi. p.233.
10. Sharma, R.C. (1985). Indian J. Ecol., 12(1).157.
11. Upadhyay, K.K. (2002). In: Productivity and energy flow in the Ramganga (w) river system, in Kumaun Himalaya, Ph.D. Thesis, Kumaon University, Nainital.
12. Joshi, B.D. Pathak, J.K., Singh Y.N., Bisht, R.C.S. and Joshi, N. (1993). Him. J. Env. Zool., 7(1): 60.
13. Pathani, S.S. (1995). In: Impact of changing aquatic environment on the Snow trout of Kumaun Himalya. FTR Ministry of Environment and Forest New Delhi. p. 53.
14. Pathani, S.S. Upadhyay K.K. and Joshi, S.K. (2002). Him. J. Env. Zool., 16(2) 151.

8

New Record of *Tor progeneius* (Mcclelland) from the River Alaknanda of Garhwal Himalaya

S.N. BAHUGUNA, A. RAJ, H. JOSHI,
A.K. BADONI, AND B. BISHT
Department of Zoology, H.N.B. Garhwal University, P.Box 70
Srinagar Garhwal, Uttaranchal, India - 246174

Abstract

New record of *Tor progeneius* (Mc Clelland) is a valuable addition in the fish fauna of river Alaknanda. The fish *Tor progeneius* is described here on the basis of various morpho-metric and meristic characters of fish; along with synonyms and zoogeographical distribution.

Introduction

Geographically the Alaknanda valley forms the central region of the Garhwal Himalaya. The Garhwal Himalaya is situated between the latitudes 29° 26' to 31° 28' North and longitudes 86° 6' East. The vast range of altitude, coupled with the complexities of a widely varying mountainous topography given rise to large differences of climatic conditions. River Alaknanda is a glacierfed river rises from the satopanth glacier on the north-eastern slope of chaukhambha peak.[4] Different species of fishes are abundantly habited in different snowfed rivers and streams of the state Uttaranchal former the part of Uttar Pradesh[1&2] There are a number of physical, geographical and biological parameters, which influence the distribution, and abundance of fish fauna[3 &11].

Materials and Methods

The fish collected from river Alaknanda near Srikot bridge, Srinagar Garhwal. The fresh fish was fixed in 8% formalin. Before preservation a small incision was given at abdominal region, without injuring the alimentary canal. Formalin injection was also given for better preservation and fixation of internal organs. The snout of the collected fish was kept downward and the caudal region upward in the jar to avoid damage of soft parts of the fish. The fish identification, classification and nomenclature are based on basis of standard methods.

Results and Discussion

Tor progeneius (McClelland) **Synonyms** : *Tor progeneius*[9], *Barbus (Tor) progeneius*[6], *Tor progeneius,*[12-13]

Local Name

Kali Mahseer.

Distinguishing Features

Length of head 5.9 cm; of caudal 5.0; height (depth) of the body 5.3 in the total length. Eye diameter 4.58 in the length of head and 2.16 in the length of snout. Body streamlined and slender, eyes small, mouth moderate, lips fleshy, barbels two pairs, scales large. Lateral line complete with 29 scales. Snout smooth, cheeks grayish. Belly silvery, lower fins orange, maxillary barbels longer than rostral barbels.

Fin Formula

D.13 (4/9); P.15; V.9; A. 8(3/5); C.20; L1. 29;Ltr. $4^1/_2$ 2½.

Total Length

32.3 cm.

Site of Collection

From the river Alaknanda a main tributary of river Ganga, near Srikot bridge Srinagar Garhwal, Uttaranchal. Date of collection: 04/08/2004.

Geographical Distribution

As per[9] the *Barbus progeneius* is inhabited in the rivers of Assam, India.[6,13] the *Tor progeneius* is a game or angling fish and geographical distributed in north-east Himalaya (Assam).

The detailed morphometric data has been given below :

1. Total Length-32.3 cm
2. Length of caudal peduncle-3.0
3. Height of caudal peduncle - 4.1
4. Eye diameter -1.2
5. Barbels number - 4 (two pairs)
6. Head length - 5.5
7. Standard length - 25.8
8. Body depth - 6.0
9. Height of dorsal fin - 4.2
10. Height of anal fin - 2.5
11. Height of caudal fin - 7.5
12. Length of dorsal fin - 4.9
13. Length of anal fin - 4.6
14. Length of caudal fin - 6.5
15. Lateral line - complete
16. Number of lateral line scales - 29
17. Number of lateral tr. Scales-4½ / 2½
18. Length of snout - 2.6

According to Gunther[7] the species is very similar to *Tor mosal* (Hamilton -Buchanan) McClelland[10] treated it as a junior synonym of *T. tor* (Hamilton - Buchanan) coated by[12]. The fish is very similar to *Tor mosal* but the presence of a fleshy fan shaped structure behind the upper jaw is an abnormal formation. Since it is not reported in other examples to *Tor mosal*[13], support our finding. All other available fish species of *Tor tor, Tor putitora* are having golden colour in these days in river Alaknanda while the *Tor progeneius* is having some blackish colour. This may be inherited genetic character that's why this may develop among the Tor population in this area as the Ganga river have no such connection with rivers of Assam. The second possibilities it may come from Nepal to Ganga river from where it may ascend to Uttaranchal snow fed rivers.

In Assam this fish is popularly known as Jungha, due to its distinct chin with a fleshy fan shaped structure behind the upper jaw[5]; but this structures could be an abnormal formation since it is not observed in other examples of this species[9]. Some workers[3,7] treat it as junior synonym of *Tor tor*, while some considered with *T. mosal* (Hamilton-Buhchanon). This is the only gangetic *Tor* with cheels, tubercles and the only *Tor* anywhere, which may occasionally lack completely a median lobe on the lower lip.[8] Present record of *Tor progeneius* (McClelland) from river Alaknanda, is of special interest.

Remarks

Fleshy fan shaped structure behind the upper jaw was observed. The lateral line possesses 27 scales in the present speciman along with the maximum number (27) of lateral line scales was also observed in this specimen.

Acknowledgements

The authors are grateful to ICAR, New Delhi for financial support and Prof. Asha Chandola Saklani, Head Zoology Deprtment, UGC-SAP, (COSIST, FIST) for providing necessary facilities in the department.

References

1. Badola, S.P. and Pant, M.C. (1973). *J. Zootomy.*, 14 (1): 37.
2. Badola, S.P. and Singh, H.R. (1977). *Indian J. Zootomy.*, 18. 115.
3. Badoni, A.K. (2003). In : Taxonomical studies of fish fauna in relation to physico-chemical nature of three important Alaknanda tributaries of district Tehri Garhwal. *D. Phil. Thesis.* H.N.B.G.U. Srinagar Garhwal.
4. Bose, S.C. (1972). In : *Geography of the Himalaya.*
5. Day, F. (1878). In : *The fishes of India being a natural inhabit the seas and freshwaters of India, Burma and Ceylon.* Text and Atlas in 4 parts. London xx + 778 pp. 195 pis. (Issued in 4 parts).

9
Observations on Ichthyo-fauna of District Pithoragarh - A Case Study

KUM KUM SHAH, B.P. OLI , PRATIBHA BISHT* & D.C. PANT
Department of Zoology, Govt. Post Graduate College,
Pithoragarh -262501

Abstract

A survey was carried out during 2002-2005 to evaluate the Icithyofauna of distt. Pithoragarh from Kali riverine system. This riverine system includes Kuthiyangti, Dhauli, Gori, Sarju, E.Ramgang, Panar and many small rivulets locally known as Gad viz. Thuli Gad, Rai Gad, CharmaGad, Gallati gad etc., which covers many villages of the distt. Pithoragarh.The collection of fishes was made with the help of villagers using cast net, hand net, hammering method, teep method etc. During three years study 26 species of fishes belonging to 16 genera & 4 families including 5 exotic carps were collected. It was interesting to note that the Schizothorax sp, Tor sp, Labeo sp. Barilius sp. were caught from every river & rivulet in each season while Glyptothorax sp , Psuedecheneis sp. & Glyptosternum were caught from E. Ramganga near Thal and Goriganga near Jaulgibi only in summer and rainy season. The maximum length and weight was 22" & 3.7 Kg of Schizothorax followed by Labeo sp. 20" & 2.8 Kg. The largest Mahseer caught was 15.6" & 1.9 Kg in length and weight respectively. Approximately 400 families of farmers from different villages are involved in fishing for their livelihood in the study area. In addition to that, culture fishery is flourishing in the district. Farmers for the pisciculture constructed approximately 400 small impoundments and rearing Ctenopharyngodon idella, Hypophthalmichthys molitrix and Cyprinus carpio. The annual production of these impoundments is approximately 3720 Kg/ha/year using advance technique of pisciculture.

Introduction

Uttarakhand state has two regions viz. Kumaon & Garhwal. The Kumaon region lies between Lat. 28°44'-30°49' N and Long. 78°45' -81 °5' E. The region comprises 6 distt. viz. Nainital, U.S.Nagar, Bageshwar, Almora, Champawat & Pithoragarh. The region has various microclimatic zones having rich natural water resources represented by a number of snow and spring fed rivers, rivulets, lakes & reservoirs. Among these districts of Kumaon, Pithoragarh is a remote area and holds great strategic importance as it shares its boundary internationally with Nepal on its east and with Tibet (China) on its North-East. Total population of the district is 4,62,149 (Census 2001), inhabit in 7100Km^2 geographical area having 2104 Km^2 area of forest cover (FSI 2001). The Kumaon has three major riverine systems i.e. Kali, Ramganga and Ganga. Among these Kali is the largest drainage system which covers almost whole of the district Pithoragarh. Information on the Ichthyofauna of any drainage system is of prime importance for management and conservation point of view. It is often said that the environmental factors affect the ichthyo-fauna of any region but no reliable database available for developmental strategy for capture and culture fishery in the region. It is not possible rather difficult to study the whole natural water resources of Kumaon. Hence an attempt was made to generate necessary scientific information for the ichthyo-fauna of the district Pithoragarh through this survey.

Many Ichthyologists have carried out work on distribution of fishes & hydrobiology of different aquatic ecosystems of Uttaranchal. Important contributions have been made by various workess[1-16].

For district Pithoragarh only few spasmodic studies are available for distribution and abundance of fishes[12-17]. Hence there is an urgent need to assess the ichthyofauna of this district for management point of view because day by day the ichthyofauna is declining due to illegal methods of fishing & habitat degradation.

Study Area

In present study efforts were made to collect the fishes from remote areas of various rivers & rivulets of Kali riverine system, where villagers depend upon fishing for their livelihood. River Kali originates from Kalapani at an altitude of 4266 masl. This joins with Dhauli at Tawaghat and with river Gori at Jaulgibi (Fig. 1 map of study area). On the other hand E. Ramganga originates from Ponting Glacier and joins with river Saryu at Rameshwar. Before confluencing with river E.Ramganga, river

Saryu joins with Panar, now it is known as Saryu. Saryu finally confluences with Kali at Pancheswar and known as Kali. It enters in the Tarai plains at Tanakpur, therforth called as Sharda. Among the Principal tributaries of River Kali, the KutiYankti, Dhauli, E- Ramganga, Gori, Saryu are glacial, of glacial origin.

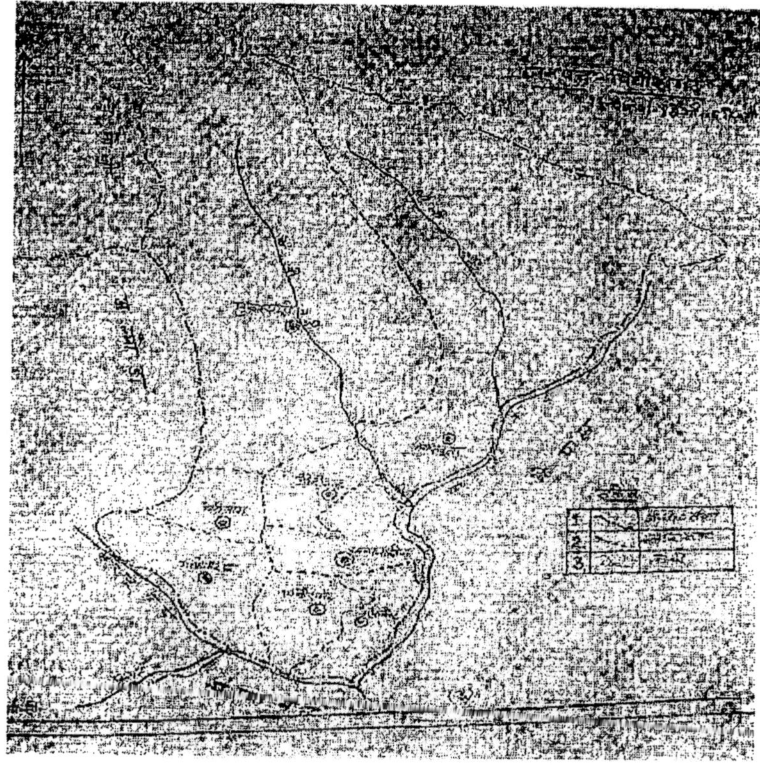

Fig. 1: Map of study area of district Pithoragarh.

Tributaries of river Kali originate in lesser Himalaya form pools, rivulets and riffles among boulders and coupled with deep gorges. Besides this, each river is connected throughout a network of spring fed small perennial rivers, brooks, streams, feeder channels and rivulets (Gad) viz. Thuli Gad, Rai gad, Galati Gad, Charmgad, Bhuj Gad, Mehar Gad, Suwalake Gad, Serali Gad, Ela Gad etc., which are major source of drinking as well as of irrigation in almost all parts of district Pithoragarh along with diversified ichthyofauna and aquatic life. Many hot water springs are also present near Dharchula & Madkot region of district Pithoragarh.

Methodology

During the study period (2002-2005) extensive survey of the river system as well as its important tributaries were conducted. The collection of fishes were made with the help of villagers using cast net, fatela jal, hand net, angling using various types of bait, viz. flour, balls, insects and earthworm etc. as well as hammering, Teep method etc. from different catching sites (Table 1). After collection, specimens were preserved in 10% formaldehyde and brought to the laboratory for identification. The identification was carried out with the help of standard methods. The hydro biological character viz. pH, D.O., temp, hardness etc. of river Kali were also estimated. In addition to the above, a survey was carried out in few stations, which are popular for their fish dishes (Macchi-Bhat, Fish Pakora & roasted fish). These places are Thal, Jaulgibi, JhulaGhat, Dharchula, Madkot, Pepali and Ghat. A survey was also made in culture fishery of the district with the help of fisheries department of the District and information with the author.

Results & Discussion

The water quality of Kali riverine system is alkaline with pH range between 8.3 -8.7 in summer & 8.0-8.3 in winter months, while water temperature ranged between 17°C -30°C in slow running rivers & rivulets and 12°C-21°C in fast running rivulets. The water is highly oxygenated with Dissolved Oxygen (DO) content 10-13 mg/l in higher Himalayan region and 7-9 mg/l in slow running water, while hardness ranged between 16.2-75.6 mg/l. The concentration of magnesium is higher due to the presence of magnesite orc in the area. Similar observations were also recorded by Raina *et.al.*[18]. In nutshell the water quality of Kali riverine system is suitable for harboring a variety of cold-water fishes.

The snow fed river Kali and its tributaries, flow with high speed among boulders between the hard rocks at high altitude. These have rocky bottom which results in occurrence of rare fishes hiding under cresices, under bowder etc. The Siluroid fishes are found in these conditions and they have developed special adhesive organs for attaching to any hard surface. So these fishes hide themselves below boulders with the help of such "specialized organs. It was observed that the catch percentage of such fishes were highest during rainy season. After rainy season the The quantum of fish haul comes down and in winters it becomes difficult to catch these Siluroid fishes. The reason behind this is that during rainy season due to furious flow of the rivers the fishes are detached from the boulders and come to the surface, so caught easily by fatela the, cast net, angling & by hammering method etc. In summer month when the water level comes down hammering method, hand picking, diversion of rivulets water etc are used to catch fishes from the rivers.

Table 1. Location wise Ichythyofauna of district Pithoragarh

Name of the River / Rivulet	*Location*	*Ichthyofauna*
1. Kalika Gad	Kalika	*Schizothorax plagiotomus. Tor tor. Nemacheilus, Berilius. Garra*
2. River Kali	Dharchula	*Labeo dyocheilus. Labeo dero. Glyptosternum. Tor tor. Tor putitora. S. plagiotomus, S. kumaonsis, Garra gotyla.*
3. Gurjee Gad	Titri village	*Pseudecheneis, Tor tor. Tor putitora, Berilius, Schizothorax.*
4. River Gori	Jaulgibi	*Labeo dyocheilus. Labeo dero, Glyptosternum, Tor putitora. Glyptothorax, Pseudecheneis.*
5. Charma Gad	Merthi	*Schizothorax.* Mahaseer. *Berilius. Nemacheilus.*
6. River Kali	Pepali	*Garra. Schizothorax.* Mahseer. *Berilius.* Labeo sp
7. River E. Ramganga	Thal	*Pseudecheneis. Tor tor. Tor putitora. Berilius.*
8. Suwalake Gad	Suwalake	*Glyptosternum. Schizothorax richardsonii, S. kumaonsis, Nemacheilus.*
9. Mahadev Gad	Nakot village	*S. kumaonsis. Ctenopharyngodon idella**. *Carassius**, *Hypophthalmichthys molitrix**. *Cyprinus carpio**.
10. Rai Gad	Dhanora, Gad	*Puntius ticto. Berilius. Schizothorax. Ophiocephalus. Nemacheilus.*
11. Thuli Gad	Cant area	Mahseer, *Schizothorax, Garra, Carassius**, *Berilius, Nemacheilus, Hypophthalmichthys molitrix**, *Cyprinus carpio**.
12. Fagali Gad	Bad Khali	*Berilius, Schizothorax.Puntius ticto, Garra, Nemacheilus.*
13. River Kali	Jhulaghat	*Schizothorax.* Mahseer, *Nemacheilus. Berilius. Garra. Labeo gonius. Ldero.*
14. River Saryu	Rameshwar	*Labeo dyocheilus, Labeo gonius. Ldero. Mastacembelus armatus. Torputitora. Tort or.* S. *richardsonii.*

* The farmers in their impoundments near Thuli gad and Nakot village rear these exotic fishes. It's a chance that these fishes have run off from these ponds during rainy season along the flowded water and thus now are found in these streams too.

Most of the streams and gad have moderate flow, gravel and sand bottom, hence provide suitable ground for the fishes present in the area. During 3 years study 26 (Table 2) species of fishes have been recorded from river Kali drainage system, belongs to 13 genera & 4 Families. Of these, family Cyprinidae is species rich (21 species) followed by Sisoridae (3 species), Channidae & Mastacembelidae (1 species of each family)

The Ichthyofauna comprises 3 sp. of *Schizothorax (S. richardsonii, S. kumaonensis, S. Plagiostomus)*, 3 sp. of *Labeo (L. dero, L. dyocheilus, L. gonius)*, 3 sp. of *Berilius (B. bendelisis, B. bola, B. vagra), 2* sp. of *Tor (T. putitora, T.tor), 2* sp. of *Garra (G. gotyl gotyla, G. annandale), 2* sp. *Nemacheilus (N. botia, N. rupicola)*, 1 sp. of *Puntius (P. ticto)*, 1 sp. each of *Glyptothorax (G. pectinopterus), Psuedecheneis (P. sulcatus), Glyptosternum (Glyptosternum sp.), Ophiocephalus (O. gachua)* & *Mastacembelus (M. armatus)*. Five species of exotic carps were also collected & these are *Carassius carassius, Ctenopharyngodon idella, Hypophthalmichthys molitrix, Cyprinus carpio communis*, and *Cyprinus carpio specularis.*

It was interesting to note that the *Schizothorax* sp, *Tor* sp, *Labeo* sp, *Berilius* sp., were found & caught from every river and rivulet in all seasons. After rainy season, due to sand mining near the riverbank by the villagers, small channels are formed, in which fingerlings of *Schizothorax* sp., *Puntius* sp. And *Berilius* sp. were found. The *Glyptothorax* sp.

The maximum length and weight was 22" & 3.7 Kg of *Schizothorax* followed by *Labeo* sp. 20" × 2.8 Kg. The largest Mahaseer caught was 15.6" × 1.9 Kg in length and weight respectively. It was observed that more than 60% captured fishes were *Schizothorax*, 10 -20% Mahseer, 10-12 % *Barilius* sp and rest others. *Labeo* species were captured in fair amount during Monsoon. Economically important fishes of the area are S. *richardsonii, S. Plagiostomus, T. putitora, T.tor, L. dero, L. dyocheilus, M. armatus, Garra* sp. & *Berilius* sp.

Ecomorphologically ichthyospecies of district Pithoragarh contain torrential (eg. *S.richardsonii, G. pectinopterus, P. sulcatus* etc.) semitorrential (eg. *L dero, T.tor, T. putitora* etc.) and very few are plain water species (eg. *Puntius* sp. etc.). Joshi *et al*[11]. described 18 sp. of fishes from Gori river system, whereas Joshi[12] described 21 sp. from Pithoragarh district including Ladhiya & Lohawati Rivers etc., which falls in district Champawat.

Table 2. Ichtyo-fauna of Distt. Pithoragarh

Order - Cypriniformes	*Family - Cyprinidae*
Genus	*Common Name*
1. *Schizothorax richardsonii*	Snow trout/ Asela
2. *Schizothorax kumaonensis*	Kumaon trout/ Asela
3. *Schizothorax plagiostomus*	Asela
4. *Tor putitora*	Golden Mahseer/ Kariya
5. *Tor tor*	Mahseer
6. *Labeo dero*	Kalabansh
7. *Labeo dyocheilus*	Kali rohu
8. *Labeo gonius*	Fogta
9. *Puntius ticto*	Ticto/ Chadu
10. *Berilius bendelisis*	Hamiltons Barila
11. *Berilius bola*	Bola
12. *Berilius vagra*	Vagra Baril
13. *Garra gotyla gotyla*	Gotyla/ Gadhele
14. *Garra annandale*	Gadhele
15. *Nemacheilus botia*	Gadera
16. *Nemacheilus rupicola*	Gadera
17. *Carassius carassius*	Crusian carp
18. *Ctenopharyngodon idella*	Grass carp } Fogta
19. *Hypophthalmichthys molitrix*	Silver carp } Fogta
20. *Cyprinus carpio communis*	Scale carp } Fogta
21. *Cyprinus carpio speculaaris*	Leather carp } Fogta

Order - Siluriformes ***Family - Sisoridae***	
Genus	*Common Name*
22. Glyptothorax pectinopterus	
23. *Pseudecheneis sulcatus*	
24. *Gyptosternum sp.*	Kabadiyal / Brahmin Machli

Order- Perciformes ***Family - Channidae***	
Genus	*Common Name*
25. *Ophiocephalus gachua*	*Snake headed fish*

Order - Mastacembeloidei Family - Mastacembelidae	*Common Name*
Genus -	
Mastacembelus armatus	***Spiny eel /Bam***

It was observed that approximately 400 families of farmers from different villages are involved in fishing for their livelihood. In addition to that culture fishery is also flourishing in the district. The culture fishery was started in 1974 in the distt., but the actual work in this field was started after two decades i.e. in 1990's. A survey was carried out and observed that Farmers have constructed approximately 400 small impoundments and rearing silver carp, grass carp & common carp in them. The approximate area of these impoundments is in between ½ nali to 4 nali (½ nali = 100m^2). The annual production of these impoundments is approximately 3720 kg/ha/yr using advance methods of pisciculture. During the year 1999, a training programme on pisciculture for fish farmers was carried out by the Dept. of Zoology Govt. P.G. college. Pithoragarh and D.R.D.A. Pithoragarh to impart basic knowledge & encourage this venture among them. Before this programme only monoculture of common carp was in practice and not giving satisfactory growth, hence farmers have not adopted this technology. To solve the problem Polyculture technique was started in 1999 by the author and a yearly calendar was prepared for Pisciculture in the hilly region of Uttarakhand.

Acknowledgement

The authors are thankful to the Principal, Prof. K.K. Bhatt, LS.M.G.P.G.C. Pithoragsfrh for providing laboratory facilities and to U.G.C. for providing financial assistance. A word of thanks is also due to Dr. B.D. Joshi, Prof. & Head Department of Zoology & Environmental Science, Gurukula Kangri Vishwavidyalaya, Hardwar; for his encouragement. The authors are also thankful to Miss Lata Bahuguna for her cordial help in identification of fish species.

References

1. Pant, M.C. (1970). *Rec. Zool. Sur. India.* 64 (1-4): 85.
2. Menon, A.G.K.(1974). In: A checklist of fishes of the Himalayan & the Indo Gangetic plain. *Bul. No.1 Inland Fish. Soc. of India. Barrackpore.*
3. Pathani, S.S. (1979). In: Studies on the ecology and biology of Kumaon Mahseer. Ph.D. thesis, K.U. Nainital.
4. Badola, S.P. (1975). *Ind. J. of Zool* 16: 57.
5. Badola, S.P. & Singh, H.R. (1981). In: *Proc. of Nat. Ac. of Sci. B.* 51(2): 133.
6. Sharma, R.C. (1984). *U.P. J. of Zool.* 4 (2): 208.

7. Rawat, J.S. (1988). In: *Kumaon Land & People*, Gyanodaya Prakashan, Nainital.
8. Singh, H.R. (1988). In: Biological studies on the Mahseer & Snow trout of Garhwal Himalaya. *FTR of ICAR Project.*
9. Bhatt S.D. & Pathak, J.K (1990). In: River pollution in India (ed. *Trivedi), Ashish Pub. House,* New Delhi.
10. Bhatt S.D. & Pathak, J.K. (1992). In: Himalayan Environment: Water quality of drainage Basins. *Shree Almora Book Depot,* Almora.
11. Joshi, B.D. & Bisht, R.C.S. (1993). *Him. Joun. Env. of Zool.*, 7: pp. 76-82.
12. Joshi, C.B. (1994). In: A Report on Cold Water Fisheries in Pithoragarh district of Kumaun Himalaya (U.P.)-Present status, Problems and Prospects. *NRC-CWF Pub. No.5A-8p.*
13. Shah, K.K. (1994). *J. Inland Fish. Soc.* 10(2):25.
14. Director, NRCCWF (1999). In: Fishes of Indian Uplands. No.2 - *NRCCWF Bhimtal.*
15. Joshi, B.D. (2000). In: *Cold water Aquaculture* (ed. Singh, H.R. & Lakra, W.S.), Narendra Pub. House, Delhi.
16. Singh, H.R. & Kumar, N. (2000). In: *Modern Trends in Biology Research* (ed. Dutta Munshi, J.S.), Narendra Pub. House, Delhi.
17. Joshi, S.N., Tripathi, G. & Tewari, H.C. (1993). In: Advances in Limnology (ed H.R.Singh).
18. Raina, H.S., Sunder, S. & Joshi, K.D. (1999). In: Resources assessment, ecological biodiversity characterstics in Kumaon Himalayas. Annual Report 1998-99, NRCCWF.

10
Fisheries and Fish Catch Composition of Kidari Jheel, Mahoba, Uttar Pradesh & Its Enhancement

B.L. PANDEY
Central Inland Fisheries Research Institute, 24, Pannalal Road, Allahabad-211 002 (U.P.)

Abstract

The fish yield of Catla catla, Labeo rohita and Cirrhinus mrigala (Indian major carps) and *Ctenopharyngodon idella* and *Cyprinus carpio* (Exotic carps) ranged from nil (1997) to 1353.60 kg (1994) - av. 360.20. Cat fishes (Wallago attu, Aorichthys aor, A. seenghala and Clarias batrachus) were fairly abundant (372.80, 1999 to 1284.00 kg, in 2000) - av. 477.7. Minor carps and miscellaneous fishes (*Labeo gonius, L. bata, Puntius ticto, Notopterus notopterus* and Murrels) were most abundnat contributing 23.64 (1994) to 88.19% (1999) towards total fish yield with an average fish catch of 1477.8 kg. The jheel was stocked with 4500 nos./ ha fingerlings (size : 70-110 mm) during 2000 with emphasis on grass carp (*Ctenopharyngodon idella* - 55.56%) and common carp (Cyprinus carpio - 27.78%). The stocking of the jheel was made with IMC (Catla 30 : Rohu 40 : Mrigala 30) only from 2001 onwards with changing of its combination.

Introduction

Jheels are most important Inland Fisheries resource of India. It can boost to inland fish production having high biological productivity in comparison to other resources. The investigation carried out by some workers pertains to Easterns and Southern states[1-5]. The previous studies showed that the fish yield from these water bodies are unimpressive due to lack of their effective stocking and scientific management. The jheels of Uttar Pradesh are unexplored for its fish yield and fish catch composition. Kidari jheel is situated between Latitude 25.18° N and Longitude 79.55° E at a distance of 20 km from Mahoba (U.P.) has been explored for its fish yield and catch composition based on the data from 1993-2002. The study definitely will help in scientific management of the jheels in Uttar Pradesh to exploite the biologically rich resource for its maximum yield.

Materials and Methods

Fish catch and stocking data has been collected from the Kidari jheel, Mahoba for the period from 1993 to 2002 with the courtesy of State Fishery Department, Mahoba, Uttar Pradesh. On the basis of data annual fish yield was estimated and interpreted. Data collected on stocking of the jheel (IMC and EC only) and 2001 onwards with IMC only in different combination were also processed and interpreted.

Results and Discussion

Kidari Jheel, Mahoba has an area of 40 ha, depth 2.0 to 10 m heavily infested with *Hydrilla* sp., *Vallisneria* sp., *Ipomea* sp., *Nelumbo* sp., and *Typha* sp. The annual fish catch ranged from 116.00 (1997) to 4199.00 kg (1998) and fish production varied between 2.90 to 104.98 kg/ha (av. 57.8 kg) during the same period, respectively (Table-1). The fish yield is very low in comparison to beels of Assam and West Bengal[1,2]. The fish yield was low during 1997, 1993 and 2002 (116.00 to 1415.90 kg) and annual producction estimated to be 2.90 to 35.40 kg/ha. Indian major carps and exotic carps *(Catla catla, Labeo rohita, cirrhinus mrigala)* (IMC) and *Ctenopharyngodon idella* and *Cyprinus carpio* (EC) ranged from nil (1997) to 1353.60 kg (1994) forming 48.81% in the total fish yield. The fishes less than 1.5 kg of third group was caught maximum during 1994 (15.26%) of the total catch while the contribution was negligible in other years. Cat fishes *(Wallago attu, Aorichthys aor, A. seenghala* and *Clarias batrachus)* were fairly abundant 14.25 (1997) to 1284.00 kg (2000) contributing 12.28 to 35.57% respectively during the same period towards total fish yield. MC and M *(Labeo gonius, L. bata, Puntius ticto, Notopterus notopterus* and Murrels) *were most abundant 101.75 (87.72%) to 3693.50* kg (87.96%) during 1997 and 1998 respectively which confirms the earlier observations in Maan and Chaur of North Bihar[4].

Table 1. Fish catch (in kg) composition during 1993-02 from Kidari Jheel, Mahoba (U.P.) (Figs, in parenthesis represent the percentage).

Year	*IMC & MC*	*Cat Fishes*	*MC&M*	*Total*	*Catch/ha*
1993	198.25 (14.00)	316.60 (22.36)	901.05 (63.64)	1415.90	35.40
1994	1356.60 (48.81)	764.15 (27.55)	655.50 (23.64)	2773.25	69.33
1995	949.00 (33.73)	297.30 (10.57)	1567.15 (55.70)	2813.45	70.34
1996	140.50 (8.97)	402.25 (25.68)	1023.50 (65.35)	1566.25	39.16
1997	—— ——	14.25 (12.28)	101.75 (87.72)	116.00	2.90
1998	7.50 ——	498.00 (0.18)	3693.50 (11.86)	4199.00	104.98
1999	16.20 (0.49)	372.80 (11.32)	2904.00 (88.19)	3293.00	82.33
2000	198.50 (5.50)	1284.00 (35.57)	2127.00 (58.93)	3609.50	90.24
2001	692.50 (24.93)	636.50 (22.91)	1449.00 (52.16)	2778.00	69.45
2002	46.00 (7.76)	191.00 (32.21)	356.00 (60.03)	593.00	14.83
Total	3602.05 (15.56)	4776.85 (20.63)	14778.45 (63.81)	23157.35	578.93
Average	360.20	477.70	1477.80	2315.70	57.89

IMC & EC (Indian major carps and exotic carps) : *Catla catla, Labeo rohita, Cirrhinus mrigala, Ctenopharyngodon idella* and *Cyprinus carpio.*

Cat Fishes : *Wallago attu, Aorichthys aor, A. seenghala, Clarias batrachus.*

MC & M (Minor carps and miscellaneous): *Labeo gonius, L. bata, Puntius ticto, Notopterus notopterus and* Murrels.

Table 2. Stocking of Kidari Jheel.

2000	1,80,000	27.78	5.00	6.60	5.00	55.56	4500
2001	50,000	—	30.00	40.00	30.00	—	1250
2002	80,000	—	40.00	30.00	30.00	—	2000
2003	1,00,000	—	30.00	30.00	40.00	—	2500

Fish Diversity

The following fishes were encountered from the Kidari jheel during 1993 - 2000.

Indian Major Carps (IMC) and Exotic Carps (EC)

Catla catla, Labeo rohita, cirrhinus mrigala, Ctenopharyngodon idella and *Cyprinus carpio.*

Cat Fishes

Wallago attu, Aorichthys aor; A. seenghala and *Clarias batrachus.*

Minor Carps (MC) and Miscellaneous (M)

Labeo gonius, L. bata, Puntius ticto, Notopterus notopterus and Murrels.

Stocking and Yield Pattern

The stocking of the jheel was made during 2000 with 1,80,000 fingerlings (size; 70-110 mm) with greater emphasis on EC *viz.* common carp *(Cyprinus carpio)* 27.78% and grass carp *(Ctenopharyngodo idella)* 55.56% and IMC 16.66% (Table-2). The stocking rate was very high 4500 nos/ha during the same period. The jheel is being stocked subsequently by IMC 1250 -2500 nos/ha/y in combination of *Catla* 30 : Rohu 40 : Mrigal 30 or changing the proportion. (Table-2). A perusal of Table-2 clearly indicates that the stocking density does not seem to have any impact in yield of IMC except 2001 where 692.5 kg of fish was harvested. This confirms the earlier observations[1,2].

Appropriate Management Strategies

Eradication of macrovegetation by mannual or biological methods by introducing herbivorous fishes. Stocking of jheels with IMC fingerlings (size: 100-150 mm) for better survival and growth and species ratio should be altered as per growth performance. Mesh size regulation should be restricted for harvesting minor carps and miscellaneous fishes. Line and bait fishery may enhance fish yield, while regular fishing by drag netting should be adopted for IMC. Mortality of IMC Juveniles during fishing operations needs to be checked, Adoption of pen culture/ cage culture for rearing of fry into fingerlings will minimize input cost of stocking. The fish yield may be enhanced upto 1000-1500 kg/ha/y by adopting effective stocking and scientific management which will contribute a lot in boosting fish production.

Acknowledgements

The author is thankful to Dr. K.K. Vaas, Director, Central Inland Fisheries Research Institute, Barrackpore (W.B.) for providing facilities and encouragement. Thanks are also due to shri R.K. Tiwari, Fishery Development officer, Goverment of Uttar Pradesh, Kirdari Fish Farm, Mahoba for providing valuable informations.

References

1. Anon (2000 a). In: Ecology and Fisheries of beels in Assam. *Cent. Inland Fish. Res. Inst.*,Barrackpore. Bull. No. 104: p. 65.
2. Anon (2000 b). In: Ecology and Fisheries of beels in West Bengal. *Cont, Inland Fish. Res. Inst.*, Barrackpore. Bull. No. 103: p. 53.
3. Chaudhuri, H. and Banerjee, S.M. (1965). In : *Misc. Contr. Cent. Int. Fish. Res. Inst.*, Barracpore, (4) : p. 29.
4. Sinha, M. and Jha, B.C. (1997). In : Cenf. *Int. Fish. Res. Inst.*, Barrackpore. Bull. No. 70: p. 64.
5. Sreenivasan, A. (1964). *Hydrobiol.*. 24 (4): 514.

11
Studies on The Status of Shell Fisheries and Their Ethno-medicinal Uses By People of Koshi River Basin of North-Bihar (India)

AMIT K. PRABHAKAR AND S.P.ROY
Department of Zoology,T.M. Bhagalpur University, Bhagalpur-812007

Abstract

This article deals about the resources of Shell Fisheries and their interaction with local people of Kosi region of North Bihar. This region is highly productive in which about 20 species of Gastropods and 10 species of Pelecypods Crabs and prawn species are dominant and exploited by the local people. Among the Gastropada, the species belonging to genera Bellamya, Pila, Lymnaea and Planorbis and among the Pelecypods species belonging to genera Lamellidens and Parreysia are recognised as edible by the aboriginals and indigenous people of the Kosi region. The pearls are being cultured for commercial purposes after the exploitation of Lamellidens and Parreysia. It was also observed that the local people of the region consumed these shell fishes to cure a number of ailments, such as rheumatism, calcium metabolism, heart diseases, conjunctivitis, giddiness, nervousness, dehydration and various gastrointestinal disorders.

Introduction

Shell fishes are natural source of high quality and easily available proteins, steroids, minerals and vitamins, poorest among poors of human beings through out the distant rural and tribal areas of country settlements. Shell fishes from economically important animal biodiversity interacting intimately with local indigenous people. Among these species belonging to *Paratelphusa, Macrobrachium, Bellamya, Pila, Achatina, Lamellidens, Novaculina and Parreysia* are eaten by aboriginal and indigenous people as well as used as medicines for the cure of a number of ailments such as rheumatism, cardiac diseases, controlling blood pressure, asthma, rickets, calcium metabolism, nervousness, giddiness besides meeting requirements of vitamins providing many shell species, fishes and crutaceans are farmed commercially and dominate world output in aquaculture resources. Among shell fish, the pearl oyster, *Unio, Pila,* blue mussels at 3.4 million tones (including shell) dominates in one hand and crustaceans, crabs *(Paratelphusa spiningera)* and prawns, *(Macrobrachium sp.)* on other hand. Considerable work has been done on the taxonomy, biology, ecology and behavior of mollusks of this subcontinent.

The objectives of the present study is to gather information on the uses of these shell fishes in the treatment of diseases most prevalent in this flood prone predominantly rural region of the country.

Materials and Methods

Study Area

North-Bihar situated at 83°21' to 88°17' NL and 24°55' to 27°31' EL, has vast lentic resources in the form of ponds, tanks, ox-bow lakes, swamps, chaurs, canals, road and railway side depressions, flood plains and these water bodies have enormous potential for fish culture. These wetlands have vast biotic potentialities in the form of plankton, macro-invertebrates, fishes, amphibians apart with avian fauna. The physico-chemical conditions of these wetlands are congenial for the proper growth and sustainable development of shell fishes. The entire Kosi basin is famous for sustainability of mollusks biodiversity and Kosi river carries high amount of alkalinity and ca^{+} ions from the foot hills of the Himalayas which are essential chemicals for the formation of shell of these animals.

Kosi basin of North-Bihar comprising districts of Khagaria, Supaul, Saharsa Madhepura and Purnea were selected for teh present study. Shell fishes exploited by different categories of people in these five districts were investigated. Extensive field trips were conducted regularly and data were collected after survey, interviews and on spot enquiries. The shell fishes were collected and preserved in 3% formalein and brought to the laboratory. The intact animals were washed thoroughly in running tap water and slightly decalcified in aqueous acidic medicine to find out growth rings. The specimens were identified with the help of available literature. The collected specimens were submitted in the museum of the University Department of Zoology, T.M. Bhagalpur University.

The uses of these shell fishes as food, medicines, vitamin supplements etc. were investigated after random sampling from the local residents of villages and block level people of the area including data on the marketing, trade and commerce of these shell fishes.

Results & Discussion

The present study on the status of shell fisheries and their ethno-medicinal uses by the inhabitants of Kosi area of North-Bihar reveals that the river basin of this region is rich in diversity of molluscan and crustacean fauna. The diversity, abundance and dominance of shell fishes indicate well established balanced ecosystem for supporting a complex food web exists in this basin. The abundance of shell fishes in terms of species diversity indicates a good life support system for fishes and birds. Shell fishes are the major component of the macro-invertebrates, they form link between zooplankton and vertebrate taxa, such as fishes and birds and play a key role in the energy flow and bio-geochemical cycle of the wetland habitats. A number of fish and avian fauna diversity directly depend upon the molluscus population in these habitats. The commercial aspect of shell fishes as raw material for food, finance, recreation, medicines, vitamins, minerals etc., for local human population and ecological aspect for increasing biological diversity and maintaining ecological balance for the animals occupying the higher trophic level of the food chain. Thus, the considerable scope with respect to the shell fisheries of medicinal value and these resources need judicious utilization on commercial basis to generate employment oppourunity and enhance the income.

Taxonomic Diversity

The present study on the status of shell fishes reveals some interesting results. A total of twenty seven species of shell fishes have been sampled and identified, which have been commercially exploited by the people of Kosi river basin of North Bihar. The taxonomic diversity in terms of number and abundance is more in Gastropod with 18 most abundant and dominant species, 7 in Pelecypoda with ecologically sensitive and commercially viable species and 2 in Crustacea with more widely used edible species among shell fishes. All these shell fishes are the chief source of food supplement and low cost of protein to middle income group of people and thus also act as intermediate food stuffs between cereals and high protein sources such as fishes and birds.

Shell fishes are found distributed amongst macro-vegetation in the wetlands. It was observed that the appreciable seasonal changes of their population many be correlated with the appearance and disappearance of macro-vegetation of the habitat. The emergent portions of the plants afford shelter to the adult stages of Mollusca. The floating vegetations constitute a biotope with varied ecological niches where almost all types of shell fishes are found. The submerged part of plants forms the habitat of-Pelecypods, and Gastropods. The free-floating vegetations such as Eichhornia crassipes afford the colonization of molluscs such as *Bellamys bengalensis, Unio marginalis, Corbicula* and others. The macrophytes provide shade, shelter and site for oviposition and development of these shell fishes.

It was observed that resource partitioning and niche specialization are the most common ecological features of the species. The members of gastropods are mainly colonized in lentic habitat, especially in the littoral zones of the river Kosi and other wetland habitats, while pelecypods were found colonized in clear waters burried in sand and they are indicative of non-polluted water and O_2 rich habitat. Thus, these shell fishes may also consider as bio-indicator of inland waters.

The species of Crustacea *Macrobrachium sp.* and *Paratelphusa spinogera* were collected frequently from the Kosi river basin during the present investigation. It was studied by Seth [14] that the river Ganga system suffered a great loss of *Macrobrachium* and this species have disappeared in the commercial fish in Ganga ecosystem. Recently in upper stretches of Kosi river basin the culture of *Macrobrachium* for food, finance and increasing ecological diversity are done by the local farmers on the line of agriculture.

Integrated shell fish-air breathing fishes-wetland birds and Makhana (*Euryale ferox*), Singhara (*Trapa bispinosa*) culture programme is successfully propagated and good income generated, a part with environmental management and life support system, propagation in the wetland habitats of this region.

Table 1. Molluscan species used for various ailmention Kosi River barin are of Bihar.

Sl. No.	*Diseases*	*Molluscan sp.*	*Method of Cure*
1.	Asthma	*Bellamya* sp.	Soup prepared from the foot of *Bellamya sp* is used to cure these diseases.
2.	Arthritis		
3.	Joint pain		
4.	Rheumatism		
5.	Conjunctivitis	*Bellamys bengalensis*	For conjunctivitis, *Bellamya bengalensis are* collected from pond and are kept in clean fresh water in a earthen pot for night and the water is used like eye drop.
6.	Rickets	*Pila sp*	Soup prepared from the eggs of *Pila* is used to cure rickets in children.
7.	Cardiac ailments	*Lamellidens sp.*	Soup prepared from the foot of *Lamellidens sp.* and *Parreysia sp.* is used to cure the cardiac ailments and blood pressure.
8.	Blood pressure	*Parreysia sp.*	Do
9.	Giddiness and dehyderation	*Lamellidens* sp.	The shell powder of *Lammellidens* sp. mixing with honey is used for the remedy of giddiness and dehydration.
10.	Nervousness	*Lamellidens sp.*	Do
11.	Night blindness	*Bellamya sp.*	Curry of the foot of *Bellamya sp* is eat en regularly by aboriginal people of Kosi region to cure night blindness and for better eye sight.
12.	Anaemia	*Paratelphusa sp. Macrobrachium sp.*	Soup and curry prepared from these crustaceans helpful in the cure of anaemia and vitamin deficiencies.

Nutritional and Medicinal Uses of Some Edible Gastropods and Pelecypods and Crustaceans

The foot of Pila sp; Viviparous sp.; Lamellidens sp; and Parreysia sp.; is large, muscular, rich in proteins, vitamins (A,B,D) and minerals, but is fat free. It is used as main nutritional element in the form of soup, curry as well as roasted by aboriginal, allied to aboriginal people of Kosi region of North Bihar. The visceral mass is also used as also used as food in curry form after elimination of gastrointestinal tract in one hand. On other hand the muscular parts and the hepato-pancreas of Paratelphusa sp. and entire muscular parts of Macrobrachium sp. are also rich in proteins, vitamins and minerals but are not fat free because these animals contain enourmous amount of fat body which are used as food.

Besides the nutritional value flesh and eggs of these species are also used as medicines by aboriginal and allied to aboriginal people of this region. Some of the medicinal uses of shell fishes are given in Table 1.

Interaction with Local People

Shell fishes play a vital role in the life and economy of the local endemic people. Bivalves and gastropods are the resources for the cottage industries, buttons, ornaments, pearls, poultry feeds, piggery feeds, lime and other resource based activities of the local people. Paddy-cum-fish-molluscs-prawn-crab cultures are the basis of aquaculture programmes of the region.

Conservation and Management of Shell Fishes

Commercial fishing of this faunal diversity has brought many species on declining trend in the habitat. The tremendous exploitation of Molluscs and Pelecypoda for food, finance, cosmetics and medicine by the aboriginal and semi-aboriginal anthropogenics has posed a great threat on the survival and propagation of these shell fishes. Recently the declining trends of shell fishes biodiversity has been greatly affected the air-breathing fishes and avian biodiversity of region. Thus conservation and management of shell fishes on the line of national heritage of aquatic ecosystem should be done along with the food chain relationship with the economically important wild life resources.

References

1. Annandale, N. (1921). *Rec. Ind. Mus;* 22:529.
2. Gupta P.O. (1969). Rec. Zool. Surv., 67(1-4): 159.
3. Hornell, J. (1917). Fish Bull., 11: 1.
4. Hornell, J. (1921). Fish Bull., 16: 10.

5. Hora, S.L. (1925). *Res. Indian Mus.* 27:401.
6. Prasad, B. (1932). Pila, the Indian Zoological Memoirs. 4.
7. Subba Rao, N.V. and Mitra, S.C. (1982). *Bull. Zool. Surv. India*. 34 (1-2): 21.
8. Subba Rao, N.Vand A. Dey. (1986). *J. Hydrobiol*, 2(3): 25.
9. Tonapi, G.T. and L. Mulherker, 1963, *J. Bombai National. Hist. Soc.*, 60(1): 104.
10. Tonapi. G. T. (1971). *J. Bom. Nat. Hist. Soc.*, 68: 115.
11. Rao, Vasishta, HS. and S.Gulati. (1971). Res. Bull, (N.S.) Punjab Univ.; 22 (3-4): 465.
12. Sharma, U.P.; Roy, S.P. and Rai, D.N. (1983). *J. Hyderobiol.*, 2 (3): 25.
13. Roy, S.P. (2003): Secondary productivity of fresh water ecosystem. In : Advances *in fish reshearch.* Vol. 3 pp. 99-108. (eds) J.S. Datta munshi, J. Ojha, T.K. Ghosh, Narendra publishing House Delhi.
14. Seth, R.N. (2002). Threatenent species of River Ganga. *Proe. National* Sy mposium on the Ecology and Biodiversity *of Aquatic Environments.* Feb. 15-17, University of Allahabad.

12

Physical Integrity of Three Selected Streams of District Nainital (Uttarakhand) in Relation to Their Fish Potential

R.K.NEGI, TARANA NEGI*, SAIF, M. AND PRAKASH CHAND

Deptt. Of Zoology & Environmental Sciences Gurukul Kangri University, Hardwar (U.A) 249404 *Govt Post Graduate College, Jind (Haryana)

Abstract

This article describes physical integrity of three selected streams of district Nainital (Uttarakhand) The width/depth ratio was maximum at Gaula stream while the entrenchment ratio was maximum at Dapka stream. On the basis of stream gradient, streams Dapka and Gaula were classified as B type and Kosi as C type. The stream subsystem was dominated by riffle-pool, cascade, plane bed, step pool in Kosi stream, Whereas it was riffle and pool at Gaula and plane bed at Dapka stream. The maximum fish diversity was reported at Kosi stream which is followed by-Gaula stream. Habitat preference was mainly restricted to shallow and deep pools and riffle.

Introduction

Structure characteristics of the lotic environment are closely associated with the occurrence of fish species. Importance of habitats and their relationship between fish are major concern to fish biologists. Habitat features have been identified as major determinants in distribution and abundance of fishes from earlier times[1] and later individual fish species as well as entire assemblage were studied for behavioral patterns in streams of North America[2-4]. Physical habitat integrity whether a necessary condition, for ecological integrity is not well defined and rarely examined in relation to the fish potential of the streams. The relation of physical habitat is quantified so as to establish better reference conditions and to document those physical habitat alterations that actually impact the fishes[5] The present study was carried out in some selected streams of district of Nainital to evaluate the physical habitat integrity in relation to their fish potential and to describe the assemblages of the species associated with the habitat type and to recognize the habitat preference of fishes.

Materials and Methods

The streams falling in the district of Nainital were studied during the months of January, 2006 to June, 2006. Samples were collected from the three selected streams viz Kosi, Gaula and Dapka at the stream reach. For the study of stream physical integrity following parameters has been undertaken. Stream depth, width, width/depth ratio, entrenchment ratio, gradient and substrate were calculated as per Rosgen[6] stream classification system. Whereas, watershed features, channel features, sediment sources, riparian vegetation and large woody debris were estimated on the spot by stream reach characterization field data sheet. Water temperature, air temperature, pH and water velocity were measured on the spot as per standard methods[7].

Biological Parameters

For the collection of fishes, cast net of mesh size 0.5 cm. was employed at different habitat of stream. Fish were collected and preserved in 10% formalin. For the identification of fishes, standard references were consulted[7,8].

Results

There was wide variation in channel width among all three streams viz. Gaula with 19.20m, Kosi, 28m and Dapka with 7m. Maximum depth and width ratio was recorded at Gaula stream and minimum at Dapka stream. In the present study entrenchment ratio was calculated as 3.0 in Dapka, 2.6 in Kosi and 1.46 at Gaula stream. On the basis of entrenchment ratio the streams under report were slightly to moderately entrenched. Gradient is very important parameter in classification of stream types. During the course of study 2% gradient was reported at Gaula stream, 1% at Kosi and 2.4% at Dapka stream. On the basis of stream gradient, Gaula and Dapka streams were placed in B type streams and Kosi as C type stream.

Channel features were unstable at Kosi and Dapka streams whereas, Kosi stream was moderately stable. The proportion of reach represented by stream morphology types was dominated by riffle, pool, run and cascade in Kosi, riffle, side pools and run at Gaula and only run at Dapka stream. This may be because of the changes in the stream substratum. In the watershed features the predominant surrounding land use was reported as forest and residential at Kosi and Dapka. Whereas, only forest was reported at Gaula stream. However, local hydrological alterations in the form of channelization and diversion of water were very much prominent at all the streams.

The erosion feature at local hill slopes was most prominent at Gaula stream which leads to the formation of large side pools at different pockets of stream reach. The degree of instream sedimentation was reported maximum at Dapka and Gaula streams as compared to Kosi stream. Composition of substrate determines the roughness of stream channel and it has a large influence on channel hydraulics. Gaula and Kosi streams were found to be dominated by large and small boulders, cobbles and pebbles, whereas, Dapka stream was dominated by pebbles. Management activities were reported in the form of extraction of gravel, sand and pebbles from Dapka stream leads to the complete loss of stream physical habitat.

The extent and width of riparian buffer zone and extent of vegetation encroachment into the stream channel was taken into consideration and it was observed that riparian buffer zone was fragmentary at all the streams. Width of riparian buffer zone was <1 channel width at Kosi and Gaula streams. Whereas, it was <5 channel width at Dapka stream. Vegetation encroachment was minimal at all the streams. Large woody debris were reported at Kosi and Gaula streams which help in the formation of pools in the streams. During the course of study shallow and deep pools were reported as the most preferable habitat for the fishes as compared to other types. Aquatic vegetation was mainly dominated by free floating and attached algae at Kosi stream, whereas, only attached algae was reported at Gaula stream and no aquatic vegetation was reported at Dapka stream. Mean water temperature was recorded 22°C at Kosi, 21°C at Gaula and 18°C at Dapka stream. The water velocity of a stream plays an important role in the existence of biotic components at any given place. In the present study maximum velocity was recorded at Kosi stream and minimum at Dapka .pH of the natural water is controlled by both geo-chemical and biological processes. During the course of study all the streams under reports were alkaline in nature.

Fish Diversity

During the course of study a total of 7 species belonging to 7 genera and one order of fishes were reported from Kosi and Gaula streams. Whereas, no fish species was reported at Dapka stream.

Fish Species Richness

From Table-3, it is clear that Kosi stream has greater fish species richness as compared to Gaula stream. When the data of fish species was calculated as average relative fish species richness, it was observed that species *Garra*

Barilius bendelisis 4.76% at Kosi stream. At Gaula stream only two species, *Barilius bendelisis* (33.33%) and *Garra gotyla gotyla* (66.67%) were reported.

Habitat Preference

Fish diversity was also calculated in the form of their habitat preference(Table2) and it was observed that species of *Barilius* and *Garra* were mainly restricted to shallow pools along the margin of stream banks where the depth of water was less. Whereas, large specimens like *Tor putitora, Labeo* and *Puntius* prefer deep pools and riffles. Fingerlings of *Barilius* species mainly prefer margins of stream where volume of water was minimum. From these results it was clear that water depth and velocity are the two major factors responsible for the distribution of fish species in different habitats of streams. It is also inferred that small fishes reported during the study. Cypriniformis mainly preferred shallow pools and riffles, whereas, large species preferred rapid, run. deep pools and cascade.

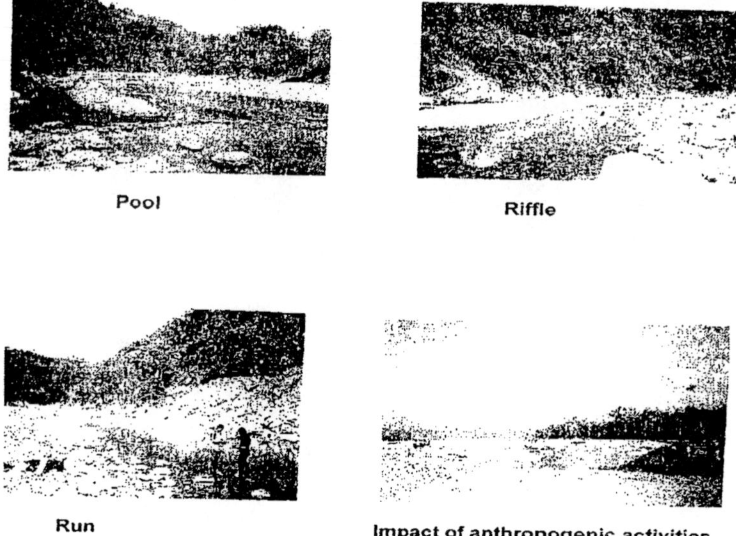

Fig. 1: An overview of the stream habitat showing pool, riffle, impact of anthropogenic activities at streambed and run.

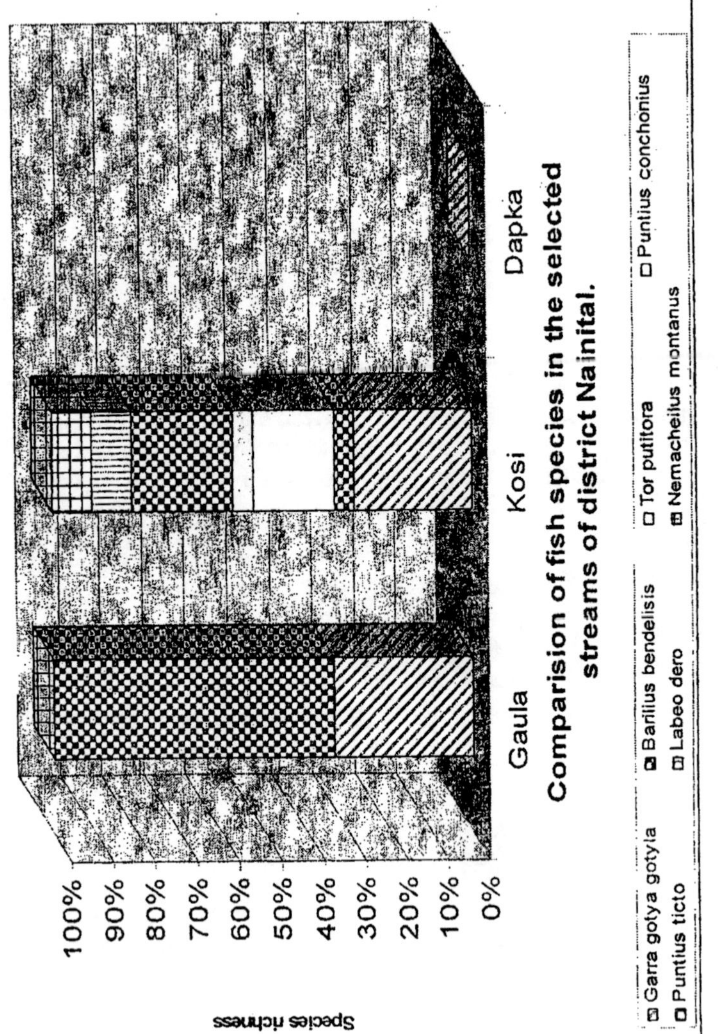

Fig. 2: Graph showing comparison of fish species in three selected streams.

Table 1. Comparative evaluation of physical habitat of three selected streams of district Nainital.

S. No.	*Physical Characters*	*Gaula(S_1)*	*Kosi (S_2)*	*Dapka (S_3)*
1.	Depth (cm)	40	60	25
2.	Width (m)	19.20	28	7
3.	Width/Depth ratio	48	46.66	28
4.	Entrenchment ratio	2.60	1.46	3.00
5.	Gradient (%)	2.0	1.0	2.4
6.	Watershed feature	Residential	Forest/Natural Residential, Diversion, Channelization	Forest/Natural Residential, Diversion, Channelization
7.	Channel feature	Moderately sable	Unstable, widening	Unstable
8.	Sediment sources	Pf, FL	Cb, Fl	Pf, Fl
9.	Substrate	B*,C*,LB,SB, C,G	C*, Sb, Pb, B	C*.SB, Pb
10.	Large woody debris	Not present	Present in channel but in some area	Not present
11.	Riparian vegetation	T,S.H.G	T,S,	T,S
12.	Aquatic vegetation	Floating alga, attached algae	Floating alga, attached algae	Attached algae
13.	Water temperature (*C)	22.0	21.0	18.0
14.	Water velocity m/sec	1.07	1.83	1.56
15.	Air temp. (°C)	26.0	24.0	24.0
16.	PH	8.0	8.0	8.3

* Dominant, SB :Small boulder, LB: Large boulder, C: Cobble, P: Pebble, T: Tree, L: Low, S: Shrub, H: Herb, G: Grass. Pf. Pool in filling. fl: Flood plain, tr: Terraces.

Table 2. Habitat preference of fishes in Kosi and Gaula streams of district Nainital.

Name of Fish species	*Pool*		*Run*	*Riffle*	*Rapid*
	Shallow	*Deep*			
Garra *gotyla gotyla*	PH	-	+	-	-
Barilius bendelisis	PH	-	+	-	-
Tor putitora	-	PH	-	+	+
Puntius conchonius	-	+	-	+	-
Puntius ticto	-	+	-	+	-
Labeo dero	-	+	-	-	-
Nemacheilus montanus	-	+	-	-	-

PH: Preferred habitat, +. Present, -: Not recorded

Table 3. Comparison offish species richness in Kosi and Gaula streams of district Nainital

Fish species	*Name of selected streams*		
	Gaula	*Kosi*	*Dapka*
Garra gotyla gotyla	33.33%	28.57%	Nil
Barilius bendelisis	66.66%	4.76%	Nil
Torputitora	Nil	19.04%	Nil
Puntius conchonius	Nil	4.76%	Nil
Puntius ticto	Nil	23.80%	Nil
Labeo dero	Nil	9.52%	Nil
Nemacheilus montanus	Nil	9.52%	Nil

Discussion

Habitat assessment of an ecoregion context provides a conceptual framework for defining biological potential[9]. Habitat variables that are believed to control aquatic biota are consequences of terrestrial features of geology, geo-morphology and vegetation that also form the basis of regionalization efforts[10-11]. Pattern of the habitat use by the fish have always been a very complex process and are not easily recognized in streams with high environmental variability[12]. In the present study, it is clear that fishes used is Garra gotya gotyla, Barilius bendelisis, Torp utitora, Puntius conchonius, Puntius ticto, Labeo dero Nemacheilus montanus different patterns of the habitat in different streams. Bratten and Berry[13] have found little evidence for discrete habitat of fish because neither individual species nor groups of fishes are associated with single habitat type.

In the case of Gaula and Kosi streams *Barilius bendelisis* was found to be very common in the side pools where depth of water was low. Johal *et al.*[14] has classified the streams on the basis of altitude range of 850-1440msl., classified the streams as A type750-1524 msl, B type,380-1150,C type,400-754 and F type 390-400. In the present study streams under report belongs to B and C types, having altitudinal range of 360-680msl.Rosgen[15] has also classified the streams on the basis of its gradient. On the basis of gradient streams were classified as B and C types, so the present findings were in line of Rosgen[15] stream classification and present work is also supported by Johal *et al.*[14]. As far as substrate is concerned B type stream was dominated by boulder, cobbles and pebbles. The present finding are inconformity with the work of Johal *et al.*[14].Vegetation cover along stream course also plays an important role in steam biodiversity. The organic material from riparian areas serves as the dominant energy source in forested stream ecosystem[16]. In the present study the streams Gaula and Kosi have rich vegetation cover hence they have good fish diversity as compared to the Dapka stream. Wang *et al.*[17] reported that the watershed connected imperviousness was the best single indicator of urbanization effects on the stream fish communities in south western Wisconsin. In the present study Dapka stream depicts the unique example of urbanization effects on fish communities. Similar observations were made by different authors from different water bodies of the world[18-19] from Washington State). Wang *et al.*[17] have also documented increased erosion, channel destabilization and widening, loss of pool habitat, excessive sedimentation, reduction in large woody debris and other types of cover as consequences of urbanization which affects the fish diversity of streams. In the present investigation, it was found that the water depth and water velocity play a very important role in distribution offish communities[14,20,27].

During the course of study numerous aggregations of minnows like *Puntius, Barilius ,Garra* were recorded in shallow pools and deep pools indicating a preferred habitat. Similar observations were made by Gorman and Karr[20] and Johal *et al.*[14] However, various species of Tor are found in deep stagnant waters[14,22]. Studies on the Western Ghats fish assemblage[23] and in other parts of the world[24-26] also found diverse groups of small fishes primarily restricted to habitats that are shallow in depth and slow in current velocity and concentrated along the stream margins in pools and riffles. Similar habitat fish community model has been reported from fish assemblage[27] where cyprinids such as *Barilius bendelisis* and *B.vagra* are common in pools and run as well as riffles and species of *Nemacheilus* in riffles. From the above observations it was inferred that stream physical habitat plays very important role in distribution of fish species as reported in Kosi and Gaula streams where maximum fish diversity was reported due to diverse physical habitats whereas, in the case of Dapka stream its physical habitat was completely uprooted which leads to absence of fish species. Hence it is concluded from above observations that stream physical habitat plays a key role in the distribution pattern of fish species.

References

1. Shelford, V.E. (1911). *Biol.Bull.* 21:9.
2. Winn, H.E. (1958). *Ecol. Monogr.* 28:155.
3. Smart, H.J. and Gee, J.H. (1979). *Can. J. Zool.* 57:2061.
4. Baker. J.A. and Ross, S.T. (1981). *Copeia.* 1981:178.
5. Rabeni, C.F. (2000). *Hydrobiologia.* 422/423:245.
6. Rosgen, D.L. (1996). In: Applied river morphology. Wildland Hydrology. Colorado. U.S.A. (Reprint Edition.).
7. Day, F. (1875-1878). In: The Fishes of India: being a natural history of the fishes known to inhabit the seas and freshwaters of India, Burma and Ceylon. Text and Atlas in 2 parts. London xx + 778,195. Reprinted in 1994 by Jagmander Book agency, New Delhi.
8. Talwar, P.K., and Jhingran, A.G. (1991). In: Island fishes. Vol. 1 & 2 Oxford and IBM Publishing Co. Pvt. Ltd. New Delhi.
9. Jayaram, K.C. (1999). In: The Freshwater Fishes of the Indian region. Narendra Publishing House, Delhi. XXVII + 551pp.+ XVIII plates.
10. Hughes, R.M.,Whitter,T.R., Rohm, C.M. and Larsen, D.P. (1990). *Envir. Manage.* 14:673.
11. Bailey, R.G. (1995). In: Description of ecoregion of the United States (2nd Ed.). Miscelleneous publication No. 1391, Map scale 1:7,500,000. U.S. Deptt. of Agr., Forest Service.
12. Omernick, J.M. (1995). In: Ecoregion: A spatial framework for environment in David, W.S. and Simon, T.P. (Eds.). Biological Assessment and Critaria: Tools for Water Resource Planning and Decision Making. CRC Press, FL.
13. Angermier. P.L. and Schlosser. J.J. (1989). Ecology. 70:1450.
14. Bratten. P.J. and Berry, C.R. (Jr.) (1997). Ecol. 12(3):477.
15. Johal, M.S., Tandon, K.K., Tyor, A.K. and Rawal, Y.K. (2002). *Pol. J. Ecol.* 50(1):45.
16. Cummins, K.W. (1974). *Ecol. Monogr.* 34:279.
17. Wang. L., Lyons. J., Kanehl. P., Bannerman. R. and Emmons, E. (2000). *J. Amer. Water Resour. Asso.* 36(5): 1173.
18. Klein, R.D. (1979). *Water Resource Bullitin.* 15(4):948.
19. Booth, D.B. and Jackson, C.R. (1997). *J. of American Analyses.* North Holand. Amsterdam.
20. German, O.T. and Karr, J.R. (1978). *Ecology.* 59:507.

21. Moyle. P.B. and Senanayake, F.R. (1984). *J. Zool. Soc. Lond.* 202:195.
22. Sehgal, K.L. (1999). In: Coldwater fish and fishes in the Indian Himalayas. River and streams. FAO Fisheries Technical Paper Rome. 385:41.
23. Arunachalam, M. (2000). *Hydrobiologia*. 430:1.
24. Finger, T.R.(1982). *J. Fresh Wat. Ecol.* 1:343.
25. Schlosser, I.J.(1985). *Ecology.* 66:1484.
26. Bain, M.B., Finn, J.T. and Brooke, H.E. (1988). *Ecology.* 69:382.
27. Edds, D.R. (1993). *Copeia*. 48.

13

Biology and Morphometric Studies of Lemon Butterfly, *Papilio demoleus* Linn. (Lepidoptera: Papilionidae) A Serious Pest of *Citrus reticulata* Var. Kinnow in Jammu

J.S. TARA AND MONIKA CHHETRY
Department of Zoology, University of Jammu, Jammu, (J&K)

Abstract

The life cycle of lemon butterfly was studied under laboratory conditions on Kinnow during post-monsoon period of 2005. Female laid eggs singly or in small groups of 2-5 mostly on under surface of tender leaves, tender twigs and sometimes on dorsal surface of tender leaves. Average incubation period recorded was 2.66+0.42 days. Average duration of first, second, third, fourth and fifth instar larvae were observed as 3.08+0.73, 2.58+ 0.86, 2.91+0.37, 3.91+0.80 and 5.08+0.80 days respectively. Pre-pupation and pupation lasted for 1.16+0.12 and 8.75±0.50 days respectively. Total life cycle duration of Papilio demoleus varied from 24.00 to 35.88 days with an average of 29.61+3.97 days. Males are smaller than females having length of 23.75+ 0.81 mm and 26.16+0.93 mm respectively, on an average.

Introduction

Citrus is considered to be one of the most important commercial crops and is cultivated throughout the tropical and sub-tropical regions of the world. In Jammu region, citrus occupies an area of 10,357 ha with an annual production of 15,154 MT of fruits (Directorate of Horticulture

Department, Jammu 2004-2005). Among the citrus varieties, Kinnow is one of the most preferentially cultivated fruit because of its high productivity and its suitability in the prevailing climatic conditions. In India about 250 spp. of insects have been reported on various citrus varieties[1]. Of these, citrus butterfly, *Papilio demoleus* Linnaeus is considered to be a major pest that causes regular heavy loss to the citrus plantation. The caterpillars feed voraciously and cause extensive damage to the leaves leaving behind midribs only. Severe infestation resulted in entire defoliation of the tree[2] thereby causing retarded plant growth and thus decreases the fruit production[3].

Materials and Methods

The biology of Lemon butterfly, *Papilio demoleus* L. was studied during post-monsoon period under laboratory conditions by the investigators in Jammu region. The collected eggs were transferred to petriplates containing fresh leaves. Leaves were changed after every 24 hrs. Moistened cotton was placed in each petriplate above which filter paper was kept to prevent the drying of leaves. First and second instar larvae were reared in small plastic bottles having perforated lids for continuous supply of oxygen. Third and fourth instar larvae were reared in covered petriplates. Fifth instars were reared in large size petriplates and were kept uncovered. The petriplates were kept inside the rearing cages for their easy pupation. The adults were provided with 10% sugar solution in cotton swabs. Observations regarding their development with respect to colour, size and the duration of different instars, pre-pupal and pupal stages were recorded daily. The morphometric measurements were recorded using standard graphic paper method. The data gathered during the experiment has been analyzed statistically and presented in Tables 1 & 2.

Results and Discussion

Mating of butterflies was observed to take place during flight. The butterflies preferred long flights and the mating was observed mostly during the morning hours. Similar observations were reported earlier[4,5].

Adult females laid eggs singly or in small groups of 2-5. Eggs are generally laid on the under surface of tender leaves and also on tender twigs by curling its abdomen. However present investigators had also observed the females to oviposit on the upper surface of leaves.

Table 1. Duration of different stages in the life cycle of Papilio demoleus L. on Kinnow.

Stage	*Duration in Days*		*Mean*	*SD*	*SE*	*cov%*
	Min.	*Max.*				
Incubation Period	2.00	2.88	2.66	0.42	0.17	13.90
1st Instar	2.00	4.00	3.08	0.73	0.30	21.42
2nd Instar	1.50	4.00	2.58	0.86	0.35	30.23
3nd Instar	2.50	3.50	2.91	0.37	0.15	11.34
4th Instar	3.00	5.00	3.91	0.80	0.32	18.41
5th Instar	4.00	6.00	5.08	0.80	0.32	14.37
Pre pupation	1.00	1.25	1.16	0.12	0.05	8.62
Pupation	8.00	9.25	8.75	0.50	0.20	5.60
Lifecycle (Egg to Adult)	24.00	35.88	29.61	3.97	1.62	8.24

Mean often samples; SD: Standard Deviation; SE: Standard Error; COV %: Coefficient of Variance

Eggs

The freshly laid eggs were smooth, spherical and creamy yellow. Sometimes unfertilized eggs were also laid that could be differentiated by its white colour and absence of black spot at the centre. The fertilized eggs later turned brown, became dark grey prior to hatching. Brar and Rataul[6] and Bhutani[7] reported the same observations as far as colour of the egg is concerned. Average diameter of egg was 0.99±0.11 mm. Ganguli and Ghosh[4] observed a slight variation in size showing 1.1mm diameter. Average incubation period was 2.66±0.42 days showing a slight variation with the findings of Maheshwarababu[8] who reported the incubation period of 2.96 days whereas Radke and Kandalkar[5] observed the incubation period of 5 days and Atluri *et al.*[9] recorded the incubation period as 4.5 days. The variations in incubation period might be due to fluctuations in the weather conditions of different regions.

First Instar Larva

Newly hatched larva after consuming its empty shell and started feeding on the tender leaves. The larva was spiny, cylindrical and brown to brownish black. It had a conspicuous creamish brown markings on its dorso-lateral sides of fifth and sixth abdominal segments which later turned white closely resembling the droppings of birds. Thorax little broader than the abdomen, head provided with two small processes. Four rows of tubercles were present on each segment of the body; two rows situated dorsolaterally and other two laterally. Three pairs of thoracic legs, four pairs of prolegs each on third to sixth abdominal segments and a pair on the anal segment was present. The first instar length varies from 2.5mm to 4.8mm with an average of 3.35 ± 0.84mm. Average duration of 1st instar larva was observed as 3.08 ± 0.73 days. Whereas some sort of similar findings with regard to its average length of 3.44mm and duration of 3.7 days has already been made by Radhke and Kandalkar[5]. However observations made by Ganguli and Ghosh[4] i.e. 2.5mm as average length and 2.64 days as average larval duration are slightly variant.

Second Instar Larva

The second instar larva became dark brown and less spiny and conspicuous creamish brown marking then turned to dirty white running obliquely along the lateral sides of the abdomen from the dorsal side but did not extend along the ventral. The average length of the second instar larva was found to be 7.68 ± 1.95mm with an average duration of 2.58 ± 0.86 days which are almost similar with the observations of Ganguli and Ghosh[4].

Third Instar Larva

The third instar larva resembled the second instar except that the length increased to 14.71±5.14mm. The third instar stage lasted on an average for 2.91±0.37 days with a range of 2.5 to 3.5 days. Rao *et al.*[10] also observed more or less similar trends in size and larval duration. These results were differing with the results of Ganguli and Ghosh[4] who reported the average length of 10mm and duration of 3.43 days.

Fourth Instar Larva

The general body colour turned blackish brown. White patch formed a V- shaped mark on the dorsum running from 2nd to 6th abdominal segments. Pair of white band became clear laterally from 7th to last abdominal segments. In the area of the white marking, paired dorsal tubercles were very minute. No dorsal horn was seen on the 10th abdominal segment. Average length of fourth instar larva was observed as 19.41 ±3.34mm with an average duration of 3.91 ± 0.32 days. These findings of the investigators are in agreement with the reports of Ganguli and Ghosh[4] which differ with the observations of Rao *et al.*[10] who reported the length as 41.43mm and the larval duration of 3.37 days.

Fifth Instar Larva

A dramatic change in colour from brown to green was seen in the final instar with mild variations from yellowish green, parrot green to whitish green. Continuous white marking was present on the ventrolateral sides of the abdominal segments. Four orange spots were present dorsally on the greyish to brownish line joining the two lateral eye spots located on the anterior side of the third thoracic segment. The average length of fifth instar larva was found to be 35.15±6.30mm with an average duration of 5.08±0.86 days. These descriptions were in agreement with the findings of various workers[4,5,10].

Thus the total larval period of citrus butterfly observed in Jammu region varied from 13 to 22.5 days with an average of 17.56±3.56 days. The length and width of head capsule of first to fifth instar larvae ranged between 0.63 and 0.71mm of minimum to 3.00 and 3.18mm of maximum respectively.

Table 2. Morphometric measurements of Papilio demoleus.

Stage	*Duration in Days*		*Mean*	*SD*	*SE*	*Cov%*
	Min.	*Max.*				
Egg	0.90	1.10	0.99	0.11	0.04	7.07
1st Instar	2.50	4.80	3.35	0.84	0.34	22.60
2nd Instar	5.00	10.50	7.68	1.95	0.79	72.80
3nd Instar	8.16	23.0	14.71	5.14	2.10	16.58
4th Instar	16.00	25.30	19.41	3.34	1.36	12.57
5th Instar	27.30	45.00	35.15	60.30	3.57	6.94
Pre Pupa	27.00	38.00	31.75	4.59	1.88	13.19
Pupa	25.00	32.00	28.66	2.44	0.98	8.53
Adult Male						
Length	21.00	26.00	23.75	1.99	0.81	7.6
Wingspan	72.50	85.00	77.17	4.82	1.97	2 2.58
Adult Female						
Length	24.00	30.00	26.16	2.29	0.93	8.00
Wingspan	85.00	95.00	88.75	4.05	1.65	4.30

A prong shaped crimson red retractile scent organ called as Osmateria was present as a defensive organ on the first segment behind the head. During the experiment the authors observed a peculiar behaviour of first to fourth instar larva which when slightly disturbed immediately raised their anterior part of the body like the hood of a snake perhaps which is presumed to be a defensive mechanism. The larvae were voracious feeders on citrus leaves leaving behind only midrib and thus causing extensive damage to nurseries and young seedlings.

Prepupa

After feeding voraciously and attaining full length the mature catterpillar shrunk in size and suspended itself in pupation by its hinder end with the help of a silken thread and enter the prepupal stage. Also same observation was made by Ayyar [11] and Atwal[12]. Prepupal stage lasted for 1.00 to 1.25 days with an average of 1.16 ± 0.12 days which measured about 31.75 ± 4.59mm in length. Rao *et al.*[10] made somewhat similar findings of 26.85mm length and prepupation stage of 1.20 days.

Pupa

Pupa showed colour variation from green, straw to grayish brown. The green pupa was reported to be rare. The average length of pupa was observed as 28.66 ± 2.44mm with an average pupation period of 8.75 ± 0.50 days. Ganguli and Ghosh[4] recorded this period of 9.24 days whereas Radhke and Kandalkar[5] observed the pupal duration of 9.15 days during their study period from Aug.-Sep. However, Atwal[12] reported the pupal period of 143 days during winter. Brar and Rataul[6] observed the pupation period of 123 ± 18.06 days during Nov.-March.

Adult

Adult butterfly was beautiful with black and yellow distinctive pattern. Head, thorax and abdomen was black dorsally but with yellow and black streaks on ventral and lateral sides of the body. Outer margin of forewings has two rows of irregular yellow spots. Blue eyespot present on the anterior angle of the hind wing. A brick red marking with a black dot outlined with a blue line anteriorly was seen on the posterior angle of the hind wing towards the abdomen. No sexual dimorphism was observed. The adult male butterfly with the body length of 23.75 ± 0.81mm and wing expanse of 77.17 ± 1.97mm whereas the female had an average length of 26.16 ± 0.93mm with the wing expanse of 88.75 ± 1.65mm.

Acknowledgement

The authors are thankful to Head, Department of Zoology, University of Jammu, Jammu for providing research facilities in the department.

References

1. Pruthi, H. And Mani, M.S. (1945). Imperial Council of Agrl. Research Science. Monograph 16: 42.
2. Bhutani, D.K. and Jotwani, M.G. (1975). Pesticides. 9(4): 139.
3. Pruthi, H. (1969). In: *Text book of Agricultural Entomology.* Indian Council of Agricultural Research: New Delhi, p. 634.
4. Ganguli, R.N. and Ghosh, M. R. (1967). Indian Agriculturist. 11: 13.
5. Radhke, S.G. and Kandalkar, H.G. (1988). P.K.V. Research Journal. 13 (2): 176.
6. Brar, T.S. and Rataul, H.S. (1973). *Punjab Hort. J.*, 13: (2 & 3): 208.
7. Bhutani, D.K. (1979). *Pesticides.* 13(4): 15.
8. Maheshwarababu, P. (1988). In: Biology and chemical control of citrus butterfly *Papilio demoleus* Linnaeus (Lepidoptera: Papilionidae) *M.Sc., (Ag.) Thesis,* Acharya N.G. Ranga Agricultural University, Hyderabad, p. 117.
9. Atluri, J.B., S.P.V. Ramana, and C.S. Reddi. (2002). *J. Natl. Taiwan Mus.* 55: 27.
10. Rao, R., Murali Krishna, A.T., Rao, S.V.R.K., Reddy B.C., Hari Babu, K. and Gopal, K. (2004). Entomon 29(3): 227.
11. Ayyar, T.V.R. (1963). In: *Hand Book of Economic Entomology For South India,* Government Press: Madras, 280.
12. Atwal, A.S. (1964). *Punjab Horticultural Journal.* 4(1): 40.

14

Studies on Some Lepidopteran Pests of Guava, *Psidium guajava* L. in Jammu Region of J&K State

TARA, J.S. AND RITA SHARMA
Department of Zoology, University of Jammu,
Jammu Tawi-180006, India.

Abstract

Lepidopteran pests are a major threat to the horticulturists especially in cases where these pests are the tissue borers. In the plains of Jammu region of J&K state, a survey conducted during 2004-2005 on Guava, *Psidium guajava* L. reveals the presence of Lepidopteran pests as borers *Deudorix isocrates* Fabricius, *Dichocroscis punctiferalis* Guenee as well as defoliators *Metanastria recta* Walker and *Spodoptera litura* Fabricius

Introduction

Guava is one of the most promising fruit crops of India and is considered to be an acquisitive, nutritionally valuable and remunerative crop. It excels most other fruit trees in productivity, hardiness, adaptability and Vit. C content. Besides its high nutritive value, it bears heavy crop every year and gives handsome economic returns. In recent years, guava is getting popularity in the international trade due to its nutritional content and processed products. However, the greatest handicaps in the large scale productivity of guava fruits are the insect pests among which the lepidopterans play an important role. Present work was thus initiated in the Jammu region of J&K state to study the mode and nature of damage caused by some lepidopteran pests to guava.

Materials and Methods

For carrying out the investigations, four fixed stations were set up in the study area. Besides, random surveys were also made on monthly basis. For the lepidopteran fruit borers of guava, studies were undertaken by bringing eggs as well as infested fruits from the field area to the laboratory where the insects were reared in the rearing cages. The larvae were provided with fresh fruits on alternate days and the adults were provided with the fruit juice and sugar syrup (1:20 dilution). For assessing the nature of damage caused by the defoliators, eggs were collected from the field and further rearing was done in the laboratory in the rearing cages. Twigs with fresh leaves were provided to the larvae on alternate days till the completion of the larval period.

Results and Discussion

The major lepidopteran pests are either fruit borers or defoliators. Amongst the fruit borers, studies were conducted for *Deudorix isocrates* Fabricius and *Dichocroscis punctiferalis* Guenee and amongst the defoliators were studied *Metanastria recta* Walker and *Spodoptera litura* Fabricius.

Deudorix Isocrates Fabricius

Common Name

Anar butterfly/Common guava blue

Family

Lycaenidae

Distribution

It is widespread and common from India to Srilanka and Burma. It frequents the plains and occurs to an altitude of 2000m. It is abundant in open plains where rainfall is low. In India, it is common everywhere except for deserts[1]. During the investigations, the pest was recorded throughout the study area.

Host Plants

Deudorix isocrates has been recorded as a pest of Guava, Orange, Mandarin, Lemon, Peach, Pear, Wood apple, Plum, Litchi, Chiku, Loquat, Anola, Mulberry, Pomegranate, etc. by various workers. The present authors have recorded it as a fruit borer of Pomegranate and Mandarin besides Guava.

Mode and Nature of Damage

The attack of this pest occurs in both the fruiting seasons i.e. winter as well as summer but very less no. of instances of its occurrence were recorded in the summer crop of guava as compared to the winter crop where approximately 8% infestation was recorded. The female butterflies (Fig. 1) lay shiny white eggs singly at variable sites (on fruit surface, at the pedicellar end, inside the calyx and even on the leaves in the close vicinity) while female butterflies to lay eggs on Pomegranate flowers also[2]. The larvae bores into the fruit (Fig. 2) and feed on the pulp and seeds making the fruit hollow from inside. The entry and the exit holes of the larvae also pave way for secondary infections by different pathogens. Pupation usually took place inside the fruit but was also recorded on the fruit surface, on the twig or any other dry surface.

***Dichocroscis Punctiferalis* Guenee**

Common Name

Castor capsule borer

Family

Pyralidae

Distribution

It is found in India, Burma, China, Srilanka, Japan, Malaysia, Indonesia and Australia.

Host Plants

Besides its main host Castor, it has been reported as a pest of some other crops like Ginger, Mulberry, Turmeric, Citrus, Jackfruit, Mango, Peach, Pear and Plum [3]. The present authors have reported it to be a pest of guava in the study area. Similar record has been made by various workers[4-8] at different places.

Mode and Nature of Damage

This pest (Fig. 3) attacks mainly the summer crop of guava in the study area. Ansari[4] reported this pest causing serious infestation to guava plantations in Punjab both in the winter and the summer crops whereas it to infest the guava crop seriously in Punjab (Ludhiana) only during the winter months[9]. The larvae of this moth bore fruits mainly but sometimes they also bore the buds. The fruits get deformed at the point of entry of the larva. The mature larva is pinkish with black spots on the body. The larvae feed on the pulp and the seeds of the developing fruits resulting in the premature dropping of the fruits. The pupa is formed in the pupal chamber usually made at the basal end of the fruit. Damage can be spotted out by the presence of frass (minute, round black pellets) coming out of the frass holes (Fig. 4).

Metanastria Recta Walker

Common name

Jamun lappet moth

Family

Lasiocampidae

Distribution

It is commonly found in India and Srilanka.

Host Plants

It is a polyphagous defoliator pest recorded from Cashew nut, Country almond, Jamun and Sapota[3]. Sandhu and Sohi[6] made the first record of its occurrence on guava in India. The present authors have recorded it as a defoliator of guava in the study area.

Mode and Nature of Damage

Spherical, dull white eggs are laid by the females (Fig.5) in clusters and are glued to the stem and branches. Full grown caterpillars are stout, cylindrical and grayish to dark brown with hair of different colours and sizes. Caterpillars feed gregariously and voraciously at night and remain crowded on the bark of the trees during day time. Contrary to this, Sandhu and Sohi[6] reported the caterpillars to feed from dusk to dawn.

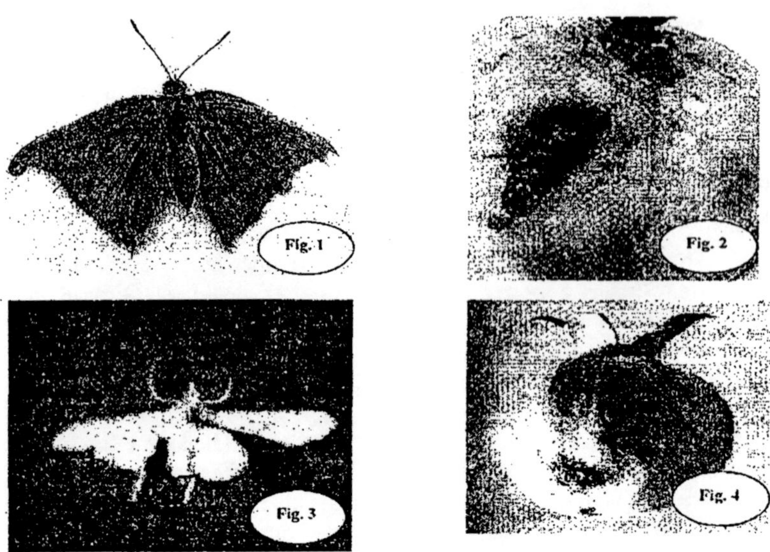
Fig. 1 Fig. 2 Fig. 3 Fig. 4

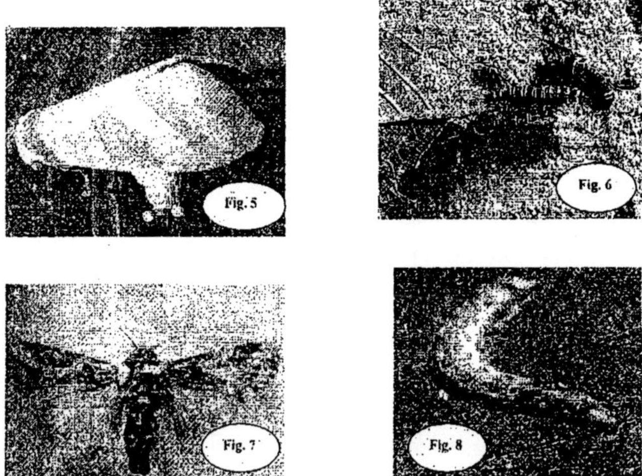

The larvae feed on the green leaves except the midrib and the large veins (Fig. 6). Consemblage behaviour and dirty grayish body matching with the bark helps them in camouflaging during day. Pupation took place in the congregated leaves. In the field, damage can be spotted out by the presence of rectangular ridged black excretal pellets fallen under the infested trees.

Spodoptera Litura Fabricius

Common Name

Tobacco caterpillar / Oriental leaf worm moth

Family

Noctuidae

Distribution

The distribution range of this pest extends from India to South East Asia and Australia i.e. throughout the tropical and sub tropical parts of the world and is widespread in India.

Host Plants

It is a polyphagous pest feeding on a wide range of native and cultivated plants. The authors have reported it to be a defoliator of guava during summer season. Similar observation was made by Das *et al.*[10] who for the first time recorded this pest from guava plantation (cultivar-Allahabad Safeda) in barwani area of West Nimar district of M.P but during Rabi season.

Mode and Nature of Damage

Eggs are laid by the females in clusters on the undersurface of leaves. The female (Fig. 7) after laying the eggs, deposits silky hair around the egg mass. Tara[11] studied the egg laying pattern of *Spodoptera litura* in Mulberry and has reported the egg laying usually in two rows but no such observation was made by the present authors during the study period. Immediately after hatching, the larvae start feeding on the mesophyll of the leaves in aggregation. The greenish brown larvae (Fig. 8) feed on the young leaves and apical soft parts of the plant mostly during early mornings or during night. The same feeding pattern was also reported in Mulberry.

Acknowledgements

The authors are thankful to the Head, Department of Zoology, University of Jammu, Jammu for providing the research facilities.

References

1. Wynter Blyth, M.A. (1957). *J. Bombay Nat. Hist. Soc.* p. 357.
2. Chander, Jagdish. (2003). In: Investigations on major insect pests of Pomegranate *(Punica granatum* L.) in Jammu forests. Thesis submitted for Ph.D. to the University of Jammu.
3. Butani, O.K. (1979). In: Insects and fruits. pp. 13-14, 19.
4. Ansari, M.A.R. (1945). *Indian. J. Ent.*, 7:241.
5. Wadhi, S.R. and H.N. Batra. (1964). In: Entomology in India. p. 239.
6. Sandhu, G.S. and A.S. Sohi. (1980). *Indian J. Ent.*, 42 (3-4): 533.
7. Panwar, V.P.S. (1995). In: Agricultural insect pests of crops and their control. p. 206.
8. Kesar, Y.K. (2001). In: Seasonal incidence and management of Lepidopteran fruit borers on Guava *(Psidium guajava* L.). Dissertation submitted for M.Sc. to SKAUST, Jammu.
9. Sandhu, G.S., I.S. Deol and A.S. Sohi. (1979). *Punjab Hort. J.* 19 (1-4):171.
10. Das, S.B., N.P. Katiyar and O.P. Verma. (1998). *Insect Environment*, 4 (1): 17.
11. Tara, J.S. (1983). In: Investigations on the insect pests of Mulberry (*Mows spp.*) in Jammu region of J & K state. Ph.D. Thesis submitted to the University of Jammu.

15

Identification of Forest Dependent Endangered Avian Habitats: Using Remote Sensing and GIS

MANOJ KUKREJA*, B.D. JOSHI AND V.K. SRIVASTAVA*****

*Rolta India Ltd. Mumbai, **Gurukul Kangri University, Haridwar ***National Remote Sensing Agency, Hyderabad

Abstract

Forested landscape in not meant for the animals only birds also inhabit the forest. Many components of the environment, including vegetation structure, plant species composition, and vegetation stratification affect the distribution of bird species. Birds breeding behaviors in an area are governed by ultimate and proximate factors. Keys to these factors are vegetations structure. Over the last few past decades the bird species population is declining continuously. At present more than hundred bird are either endemic or endangered in India alone. Long-term changes in bird populations occur in response to environmental change. The present study attempts to analyze the mode and magnitude of distribution of few birds preferred tree species. Density distribution of such trees in a forested landscape has been utilized to map the habitat for birds using remote sensing data.

Introduction

Each species of bird breed and spends its non-breeding season, respectively in a particular part of the earth-often the same - which from the geographic range or distribution of that species. In these areas, birds occupy certain habitats and have certain characteristic behaviors, interacting with other birds, food sources, predators and other plants and animals, which share their habitat. Birds are important environmental sentinels and changes in their population often reflect man's impact on the environment. Birds are very emotive subjects being fairly visible and attractive.

However, indiscriminate removal of forests has extensively damaged the bird's habitat, affecting the variety and variability in bird's population. Tribal population living in and around the sanctuary cut trees to meet their needs, thereby unknowingly destroying the habitat — the bird's richness and their abundance. More than 290 varieties of birds have been sighted in Kolhapur district, India including the study area i.e. Radha Nagari Sanctuary. These birds include the small woodpeckers to large vultures. Some of these birds are carnivorous that feed on the insects and the Caracas of the animals. Some are omnivorous feeding on both plants and animals some are herbivores feeding only on the plants.

Understanding bird's species requirements allows us to determine potential reasons for population declines and to formulate management recommendations that enable us to maintain viable populations. Miller *et al.*[1] have developed habitat suitability model to increase the knowledge of avians distribution, while Loehle *et al.*[2] have developed a landscape analysis method incorporating bird habitat model to schedule timber harvesting, without adversely effecting the bird population.

Many components of the environment, including the mode and magnitude of vegetation affect the distribution of bird species. Keys to these ultimate factors, such as food availability for nestlings, are perceived in advance through proximate factors -aspects of the physical habitat, especially vegetation structure. Ornithologists suspect that forest fragmentation harm birds by increasing their susceptibility to predation and nest parasitism.

The focus of this study is to understand the tree dynamics in relation to the need of variety forest bird species that belong to numerous taxonomic groups, including Hornbills, Flycatcher and Pigeon.

Two varieties of Hornbill birds i.e. Great Pied Hornbill and Malabar Pied Hornbill are the endangered variety, and White bellied blue Fly Catcher Nilgiri wood Pigeon are endemic to the region.

Great Pied Hornbill

The *Great Pied Hornbill, Buceros bicornis* (130 cm) is a huge pied bird with a large casqued, yellow bill. It has broad white-wing bars, white tips and white tail with a black central bar. The vent is white, as are the things. The face, belly and lower breast are black. The upper breast and neck are yellowish. The huge yellow casque extends from the rear crown to where the thick bill de-curves. The sexes are similar though the female is smaller. Makes a noisy deep barking call while the wings make a loud whooshing sound, often audible before the bird is seen. Feeds on tree - top fruits, which it tosses in the air before swallowing. Great pied Hornbill is herbivopes bird feeds primarily on wild figs, nutmegs and drops of various trees. It breeds during February to April. The Great Hornbill is endangered mostly due to deforestation. The habitat is being lost so rapidly that the hornbill has no place to go, and can't adapt to new habitat quick enough. They are also hunted for their flesh, feathers, and casque.

Malabar Pied Hornbill

The *Malabar Pied Hornbill Antracoceros coronatus,* similar to Great Pied hornbill, has broad white outer tail feathers and the white wing tips too are broader. The casue is longer and largely black and the throat patch is pink. The calls consist of various high-pitched cackles and squeals and also a fast ka ka ka ka. It inhabits open forests and groves. Malabar pied Hornbill also feeds on the figs, drupes and berries, having the breeding season between March to April. Analysis of feeding and breeding and behavior of these birds led to the identification of thirteen tree species from the area that harbors these birds.

White Bellied Blue Flycatcher

The white bellied Blue Flycatcher is an endemic bird. It feeds on the berries and is found in the evergreen forest. The breeding season of this bird is from February to September.

Nilgiri Wood Pigeon

Nilgiri wood Pigeon a herbivorous bird, is also an endemic bird. This species of the bird is highly vulnerable to extinction. It feed on the fruits, berries and buds. The nesting is on the moderate size tree in Shola forest. Its breeding season extend from March to July.

Objectives

Many birds species are adversely affected by human activities at large spatial scales and their conservation requires detailed information on distribution of tree species which meets the feeding and breeding requirements of these birds. Therefore, the objectives of the present study was:

- To analyze the distribution of tree in respect to the breeding and feeding behavior of birds.
- To locate potential suitable habitat for bird species in the RWS, Kolhapur, and Maharashtra with satellite imagery and GIS analysis.
- To identify the endangered birds in the area for the conservation.

The study area, Radhanagari Wild life Sanctuary, Located in the district of Kolhapur, Maharashtra, India, is a bison sanctuary. Along with the bisons, many more animals are found. Sanctuary is locate between the two major reservoirs viz. "Shahu Sagar" and "Laxmi Sagar" in Kolhapur district. The entire sanctuary area is undulating with steep escarpments. The total area of the sanctuaries about 440 sq. km. The major forest types found in the area.

- Tropical moist deciduous forest (covers an area of 152 sq km of the sanctuary)
- Tropical semievergreen forest and (covers an area of 36.2 sq km of the sanctuary)

Tropical evergreen forest (covers an area 5.4 sq km of the sanctuary)

Materials and Methods

Satellite Data

False color composite (hard copy) on 1:50,000 scale for the area and digital data of IRS 1D LISS-III (path/row : 96/61) for the month of 18 February 2002 was used.

Ancillary Data

Given below is the list of ancillary data used in he present study.

- DCW (Digital Chart of World) - Village location, road network and drainage information available on these maps has been directly used.
- Survey of India top sheets on 1:50,000 scales.
- Forest types of India (Champion and Seth, 1968);
- Forest Management map of the area available with State Forest Departments.

- Administrative boundaries : Sanctuary boundary has been taken from the management map of the forest (scale 1: 100, 000).
- Bird Census Data (available with State Forest Department).

Field Data and Ground Truth

Phyto-sociological data was" collected from randomly distriuted, 23 samples each of 25 × 25m size during January and February 2002.

Habitat Analysis

Literature was screened to understand the ecology of the birds found in the region so as to understand the foraging and reproduction behavior of species which could lead to an idea of species interaction between different habitats, More than 290 varieties of birds have been sighted in the district that inculdes the bird sighted in and around the sanctuary. Analysis of feeding and breeding behavior of these birds led to the identification of thirteen tree species from the area, which harbor these birds. Habitat suitability study involves an analysis of the complex inter relationship among various environmental factors that exist over a geographical area.

Results and Discussion

Literature was screened to understand the ecology of these birds. However, only 93 birds could be screened. These birds were segregated in terms of their feeding and breeding habits. 33 birds were found to be herbivorous and 31 carnivorous in their feeding behavior. The remaining birds were found to be omnivore. Following 22 birds (Table 1) were found to prefer wild trees found in the sanctuary.

Forests

Amongst the forests, moist deciduous forest has wide distribution in the next dominant forest type of the region covers an area of 36.172 km^2. Evergreen forest covers an area of 5.394 km^2 and is located on the mountain tops and clifts more towards central to southern part of the region.

The trend of various forest type distribution in the landscape is clear from the land use/land cover map. The high hills and east facing slopes of the area show evergreen forest. As one goes to wards the northeastern side occurrence of moist deciduous, semi evergreen and degraded forest increase. Agricultural land, fallow land and habitation have been considered as barren land[3].

Vegetation Analysis

Table 2 gives the general characteristics of forest vegetation found in the sanctuary. It was observed that though the total number of tree species and tree density per sample was lowest in evergreen forest, but the basal area on sample basis or on tree basis and tree height was highest. Thus, the trees found in evergreen forest were well grown owing to the better differentiation of niche in this type of forest.

The moist deciduous forest contains highest number of tree species and tree density per sample. However, the growth of the trees is poor as compared to the evergreen trees as evident from basal area per tree and means height of the tree. The semi evergreen forest was more like moist deciduous in terms of the growth of the tree species as also in terms of number of species and density.

It is of interest to note diversity (calculated in terms of Shannon-Weiner index at tree level is highest in moist deciduous forest and lowest in evergreen forest. However, diversity among the shrub was found to be highest in evergreen forest, followed by semi evergreen and moist deciduous forest respectively. The diversity at herbaceous layer was again found to be highest in moist deciduous forest and lowest in evergreen* forest. The diversity at shrub level in evergreen forest may perhaps be due to regeneration of evergreen species. Because of well grown trees in this type of forest, insolation may perhaps be a limiting factor for herbaceous vegetation to grow. Hence, low diversity at Birds listed in Table 1, prefer the berries, fruits and flowers of the following 13 tree species (Table 3) as their feed and also for nesting during breeding season. These species are of wide occurrence in the dorest. Out of the mentioned 13 tree species, five were found in evergreen forest, 12 in semi evergreen and 12 in moist deciduous forest, *Terminalia tomentosa* was dominat in all three types of forest. However, its mean height was 12.3 m in evergreen forest and about 7 m in semi evergreen and moist deciduous forest and its density was almost same 11 in evergreen and 10 in the other two type of forest. *Euginea jambolana* though common in all the type, was dominant in evergreen and dominant in semi evergreen and moist deciduous forest.

Density Map

These trees are present in all the forest found in the sanctuary. However, they vary in density and frequence. *Euginea Jambolana* was the most frequent tree species that was present in almost all the sample (Frequency = 95%) *Terminalia chebula and Terminalia paniculata* had the frequency of 70% and 65% respectively. However, *Ficus tslakela, Bauhinia racemosa* and *Butea monosperma* .

Had very low frequency (Frequency = 4%). The frequency of other tree species varied between these two extremes. The density of each species was not evaluted separately. However, collective density per sample was assessed. It was observed that density of trees varied from 8 percent to 59 percent. Base on this, the satelite data was clasified to generate the density map of the sanctuary, which indicates the presence of suitable tree species and the image indicates their distribution in the sanctuary.

Habitat Suitability Map

The suitability of habitat for bird sighted in the sanctuary will depend on the availability of the preferred tree species. More abundant the trees are, more suitable is the habitat. Therefore, the density classes were ground into least suitable class (density less than 20 percent), moderately suitable class (density between 20 and 50 percent) and highly suitable class (density more than 50 per cent).

Table 1. Birds found in the sanctuary and their feeding habits (*Endemic ** Endangered).

S. No.	*Name of the bird*	*S. No.*	*Name of the bird*
1.	Malabar Grey-Hornbill	12.	Hill Myna
2.	Malabar Pied Hornbill	13.	Quaker Babbler
3.	Large Green Barbet	14.	Black Bird
4.	Crimson Breasted Barbet	15.	Lotn's Sunbird
5.	Indian great Black Woodpecker	16.	Yellow legged Green Pigeon
6.	Glodfronted Chloropisis	17.	Green Imperial Pigeon
7.	Red Wented Bulbul	18.	Blue winged Parakeet
8.	Black Bulbul	19.	Indian Lorikeet
9.	White Browed Bulbul	20.	Nilgiri wood Pigeon*
10.	Red spur fowl	21.	Great Pied Horn bill**
11.	Golden Oriole	22.	White bellied blur fly catctier*

Table 2. Vegetation characteristics in the sancturary.

Vegetation characteristics	*Forest type*		
	Moist deciduous	*Semi evergreen*	*Evergreen*
No. of tree species	21	20	9
Density per sample	98	89	30
Basal area $(m)^2$ per sample	32.65	28.19	54.60
Basal area $(m)^2$ per tree	0.443	0.318	1.828
Mean height (m)	6.9	7.0	17.5
Diversity index at tree layer	3.67	3.48	2.39
Diversity index at tree layer	2.67	3.00	3.33
Diversity index at herb layer	1.18	1.12	1.07

Table 3. Trees preferred by bird for feeding and breeding behavior.

S. N.	*Tree species*	*Forest type*
1.	*Carissa carandas*	Moist deciduous, semi evergreen
2.	*Euginea Jambolana*	Moist deciduous, semi evergreen, evergreen
3.	*Erythrina indica*	Moist deciduous, semi evergreen
4.	*Ficus glomarata*	Moist deciduous, semi evergreen
5.	*Ficus rumphii*	Moist deciduous, semi evergreen
6.	*Terminalia bellaria*	Moist deciduous, semi evergreen
7.	*Terminalia chebula*	Moist deciduous, semi evergreen, evergreen
8.	*Terminalia paniculata*	Moist deciduous, semi evergreen, evergreen
9.	*Terminalia tomentosa*	Moist deciduous, semi evergreen, evergreen
10.	*Zyziphus rugosa*	Moist deciduous, semi evergreen,
11.	*Ficus taslakela*	Evergreen
12.	*Bauhinia raccemosa*	Moist deciduous, semi evergreen
13.	*Butea monosperma*	Moist deciduous

Accordingly an other image was generated to show the distribution of habitat suitability classes and it was observed that highly suitable habitat was located towards the periphery of the forests, close to human settlements. Hence, liable to distrubances. Studies have shown that roads and their associated intrusions can inhibit forest interior bird activity by way of fragmenting the forest and loss of nich. Moderately suitable class was evenly distributed and the least suitable class contained the large part of the forest. Thus, the most suitable is more exposed to distrubances.

Acknowledgements

The authors are thankful to Dr. P.S. Roy, Project Director, DOS-DBT project and Deputy Director, NRSA, Hyderabad, for his constant encouragement and guidance during the study.

Refferences

1. Loehle, C, Paul, V.D. Wiggly, T.B., Mitchell, M.S. Rutzmoser, H.S., Aggett J., Beebe, J.A. and Smith, M.L. (2006). A method for lndscape analysis of forestry guidelines using bird habitat moders and the habplan harvest scheduler, *Forest Ecology and Management*, Vol. 232, p. 56.
2. Miller, T.A. Mulvihill, R.S. Lanzone, M.J., Broks, R.P. and Bishop, J.A. (1998). Assessment and development of bird-Habitat models using data from the 2 Pennsylvania Breeding Atlas, *Carnegie Museum of Natural History*, Pennsylvania.
3. Srivastava, V.K., Neeti, Kukreja, M.K. and Kale V. (2003). Biodiversity characterization at landscape level usingremote sensing and GIS in Radhanagari Wildlife sanctuary, Kolhapur, Final Report, SAC/ RESA/ FLPG/ FED/ DOS-DBT/ SRI 01/ 2003, Sa (ISRO), Ahmedabad. p. 60.
4. Futuriy (2004). In: Habitat analysis - Case Study. Futurity, Inc, Chicago.

16

Diversity and Status of Avifauna of District Kathua (J&K)

SANJEEV KUMAR and D.N.SAHI
Department of Zoology, University of Jammu, J&K-180006.

Abstract

The present investigation was undertaken in the district Kathua of J&K state from January 2002 to December 2005 with the main objective of studying the diversity of the avian species. The area under study i.e. district Kathua lies between 32° 23' and 32° 44' N latitudes and 75° 02' and 75° 44' E longitudes with an altitude ranging from 240m to 4200m.To carry out the work, the study area was divided into five different stations and periodic surveys were conducted to make observations. The study revealed 117 species of birds belonging to 39 families and 12 orders along with their Resident/Migrant status. Of the 117 species reported, 91 species are resident recorded through out the year, 13 species show seasonal altitudinal migrations between low and high altitudes, 10 species have been reported to be winter migrants visiting the study area during winter only and 3 species were reported to be summer migrants visiting the study area during summer only. Two species i.e. Large Green Barbet and Yellow Backed Sunbird were recorded exhibiting an extension in their distributional range. In addition to this, an effort has also been made to compile a local abundance status of the species reported.

Introduction

Birds are considered as Masters of Air and are among the most familiar animals of our environment and through the years people have observed their activity with keen interest. The studies on birds have been conducted from time to time in India and Jammu and Kashmir State also. But, so far as the knowledge of present investigator is concerned, there is no detailed study on the birds of Jammu region except a few records[1-13]. So, there is an imperative need to work out the avian diversity of this area. To make a beginning in this direction, the proposed study was undertaken in Kathua district to fill the above mentioned gap in this context.

Study Area

The district Kathua is situated in south-east of Jammu and Kashmir state and lies between 32°-23' and 32°-44 North latitudes and 75°-02and 75°-44 East longitudes. The area is characterized by vivid geographical formation extending from Lesser Himalayas in North to alluvial Chenab and Ravi in South.

Fig.1: Showing the map of district Kathua.

Climate

Kathua district experiences subtropical to temperate climate. Climatic conditions are dry sub humid to arid. The lower parts of district along national highway are characterised by hot and dry season from March to June. The upper reaches of Billawar and Basohli tehsils have comparatively cooler summers with occasional showers. There are three well defined seasons:

winter season (Mid Nov. to Feb.), summer season (May to end of June), monsoon season (July to September) with short spans of spring (March to April) winter and summer season and autumn (October to mid November) between monsoon and winter season. Average rain fall ranges from 100-120cms. About 80% of total annual rain fall comes during monsoon and rest of 20% occurs during rest of year. Humidity is high during monsoon, January is coldest and May-June are the hottest months of the year.

Methodology

In order to record the avian diversity, periodic surveys were undertaken in the study area by adopting systematic field procedures and techniques for survey and collection of different species. The study area was divided into five different stations - Hiranagar (240m), Kathua (245m), Billawar (680m), Basohli (677m) and Bani (1260m) to make a thorough survey of avifauna of district Kathua. The surveys were conducted within 20 km of selected stations from 6 am to 11 am in morning and 4:30 pm to 6:30 pm in evening during summers and 7:30 am to 11:30 am in morning and 3:30 pm to 5:30 pm in evening during winters. In addition to these fixed timings of surveys, some irregular visits were also planned and made during other hours of the day. Line transect method and point transect method were employed for making observations.

The nomenclature followed in the present work is in accordance with those given in the "Handbook of Birds of India and Pakistan[5]. For identification and field diagnosis of birds, colourful plates of Ali and Ripley[5], Ali[14], Grimmett *et al.*[15] and Grewal *et al.*[17] have proved quite helpful. Classification of birds is in accordance with Grewal *et al.*[16] followed after *An Annotated Checklist of the Birds of the Oriental Region*[17]. Tools used include binoculars (8-17 × 40X, Super Zenith) and photographic camera (Canon) with 300 X zoom.

Results and Discussion

During the course of present investigation, 117 species of birds belonging to 39 families and 12 orders were reported from district Kathua of J&K state (Table 1). Grimmett *et al.* (1998) reported about 1300 species of birds belonging to 78 families and 17 orders in India. The study revealed that out of the total bird species recorded from the study area,-order Passeriformes is the largest order represented by 57 species which constitutesabout 48,7% followed by order Cicconiformes representing about 14.5% (17sps.) of the otal species reported from the study area. Order Galliformes represents 6.8% (8 sps.), orders Coraciformes and Piciformes constitute 6.0% each (7 sps.), Columbiformes 5.1% (6 sps.), orders Psittaciformes and Cuculiformes 3.4% each (4 sps.), Gruiformes

2.5% (3 sps.), Strigiformes 1.7% (2 sps.), and orders Bucerotiformes and Upupiformes represents 0.8% each (1 sps.).

Table 1. Systematic list of Birds with their Resident/Migrant status observed in the district Kathua.

S. No.	*Name*	*Resident/ Migrant Status*	*Abundance*
	Class: AVES **Order: CICCONIFORMES** **Family: PODICIPEDIDAE**		
1.	Little Grebe *Podiceps ruficollis capensis* Salvadori	Rst	R
	Family: PHALACROCORACIIDAE		
2.	Little Cormorant *Phalacrocorax niger* (Vieillot)	Rst	R
	Family: ARDEIDAE		
3.	Indian Pond Heron *Ardeola grayii grayii* Sykes	Rst	C
4.	Night Heron *Nycticorax nycticorax* (Linnaeus)	Rst	R
5.	Little Egret *Egretta garzetta* (Linnaeus)	Rst	R
6.	Cattle Egret *Bubulcus ibis coromandus* Boddaert	Rst	C
	Family: ACCIPITRIDAE		
7.	Pariah Kite *Milvus migrans govinda* Sykes	Rst	C
8.	Black Winged Kite *Elanus caeruleus* (Latham)	Rst	O
9.	Indian Shikra *Accipiter badius dussumieri* Tern mi nek	Rst	C
10.	Steppe Eagle *Aquila nipalensis* Hodgson	WM	R
11.	Indian White Backed Vulture *Gyps bengalensis* (Gmelin)	Rst	R
12.	Indian Long Billed Vulture *Gyps indicus* G. R.Gray	Rst	O
13.	Himalayan Griffon Vulture *Gyps himalayensis* Hume	Rst	O

Contd.

14.	White Scavenger Vulture *Neophron percnopterus* (Linnaeus)	Rst	O
	Family: CHARADRIIDAE **Sub family: RECURVIROSTRINAE**		
15.	Black Winged Stilt *Himantopus himantopus himantopus* (Linnaeus)	Rst	O
	Family: SCOLOPACIDAE **Sub family: SCOLOPACINAE**		
16.	Common Sandpiper *Actitis hypoleucos* Linnaeus	WM	O
	Family: CHARADRIIDAE **Sub family: CHARADRIINAE**		
17.	Red Wattled Lapwing *Vanellus indicus indicus* (Boddaert)	Rst	C
	Order: GRUIFORMES **Family: RALLIDAE**		
18.	Indian Moorhen *Gallinula chloropus indica* Blyth	Rst	R
19.	Indian White Breasted Waterhen *Amaurornis phoenicurus phoenicurus* (Pennant)	Rst	C
	Family: GRUIDAE		
20.	Sarus Crane *Grus antigone antigone* (Linnaeus)	Rst	R
	Order: GALLIFORMES **Family: PHASIANIDAE**		
21.	Grey partridge *Francolinus pondicerianus* (Gmelin)	Rst	C
22.	Black partridge *Francolinus francolinus* (Linnaeus)	Rst	O
23.	Jungle Bush Quail *Perdicula asiatica* (Latham)	Rst	C
24.	Indian Red Jungle Fowl *Gallus gallus* (Linnaeus)	Rst	C
25.	Indian Peafowl *Pavo cristatus* Linnaeus	Rst	R
26.	Kaleej Pheasant *Lophura leucomelanos hamiltonii* (J.E. Gray)	Rst	R
27.	Chir Pheasant *Catreus wallichi* (Hardwicke)	Rst	R

Contd.

28.	Monal pheasant Lophophorus impejanus *(Latham)*	Rst	R
	Order: COLUMBIFORMES **Family: COLUMBIDAE**		
29.	Little Brown Dove S. *senegalensis cambayensis* Gmelin	Rst	C
30.	Indian Ring Dove *Streptopelia decaocto decaocto* (Frivaldszky)	Rst	C
31.	Red Turtle Dove S. *tranquebarica tranquebarica* (Hermann)	Rst	R
32.	Rufous Turtle Dove S. *orientalis meena* (Sykes)	Rst, AM	C
33.	Indian Spotted Dove S. *chinensis suratensis* (Gmelin)	Rst	C
34.	Indian Blue Rock Pigeon *Columba livia intermedia* Strickland	Rst	C
	Order: PSITTACIFORMES **Family; PSITTACIDAE**		
35.	Large Indian Parakeet *Psittacula eupatria nipalensis* (Hodgson)	Rst	O
36.	Blossom Headed Parakeet *P. cyanocephala* (Forester)	Rst, AM	O
37.	Rose Ringed Parakeet *P. krameri manillensis* (Bechstein)	Rst	C
38.	Slaty Headed Parakeet *P. himalayana* (Lesson)	Rst	O
	Order: STRIGIFORMES **Family: STRIGIDAE** **Subfamily: STRIGINAE**		
39.	Northern Spotted Owlet *Athene brama indica* Franklin	Rst	C
40.	Great Horned Owl *Bubo bubo bengalensis* (Franklin)	Rst	R
	Order: CORACIFORMES **Family: CORACIIDAE**		
41.	Blue Jay *Coracias bengalensis bengalensis* (Linnaeus)	Rst	C

Contd.

	Family; ALCEDINIDAE		
42.	Small Blue Kingfisher *Alcedo athis bengalensis* Gmelin	Rst	R
43.	White Breasted Kingfisher *Halcyon smyrnensis smyrnensis* (Linnaeus)	Rst	C
44.	Pied Kingfisher *Ceryle rudius leucomelanura* Reichenbach	Rst	O
45.	Himalayan Pied Kingfisher *Ceryle = (Megaceryle) lugubris continentalis* Hartert	Rst	O
	Family; MEROPIDAE		
46.	Indian Small Green Bee-eater *Merops orientails orientalis* Latham	Rst	C
47.	Blue tailed Bee-eater *M. phillipinus phillipinus* LinnaeusF	Rst	O
	Order: CUCULIFORMES **Family: CUCULIDAE**		
48.	Indian Koel *Eudynamys scolopacea scolopacea* (Linnaeus)	Rst	C
49.	Pied Crested Cuckoo *Clamatorjacobinus serratus* (Sparrman)	SM	O
50.	Western Sirkeer Cuckoo *Phaenicophaeus leschenaultii sirkee* (J.E Gray)	Rst	R
	Family: CENTROPODIDAE		
51.	Crow Pheasant *Centropus sinensis sinensis* (Stephens)	Rst	O
	Order: UPUPIFORMES **Family: UPUPIDAE**		
52.	European Hoopoe *Upupa epops epops* Linnaeus	Rst	C
	Order: PICIFORMES **Family: CAPITONIDAE**		
53.	Large Green Barbet *Megalaima zeylanica zeylanica* (Gmelin)	Rst	C
54.	Coppersmith *M. haemacephaia indica* (Latham)	Rst	C
55.	Blue Throated Barbet *M. asiatica* (Latham)	Rst	R

Contd.

56.	Himalayan Great Barbet *M. virens* (Boddaert)	Rst, AM	C
	Family: PICIDAE **Subfamily: PICINAE**		
57.	Lesser Golden- backed Woodpecker *Dinopium benghalense benghalense* (Linnaeus)	Rst	C
58.	Yellow Fronted Pied Woodpecker *Picioides maharattensis maharattensis* (Latham)	Rst	O
59.	Northern brown crowned pygmy woodpecker *Picoides (Dendrocopos) nanus nanus* (Vigors)	Rst	R
	Order: PASSERIFORMES **Family: LANIDAE**		
60.	Rufous-backed Shrike *Lanius schach erythronotus* (Vigors)	Rst	C
	Family: ORIOLIDAE		
61.	Indian Golden Oriole *Oriolus oriolus kundoo* (Sykes)	SM	O
	Family: DICRURIDAE		
62.	Black Drongo *Dicrurus adsimilis albirictus* (Hodgson)	Rst	C
	Family. STURNIDAE		
63.	BankMyna *A. ginginnianus* (Latham)	Rst	C
64.	Indian Myna *Acridotheres tristis tristis* (Linnaeus)	Rst	C
65.	Brahminy Myna *Sturnus pagodarum* (Gmelin)	Rst	C
66.	Indian Pied Myna *Sturnus contra contra* Linnaeus	Rst	O
67.	Northern Jungle Myna *A. fuscus fuscus* (Wagler)	Rst	R
68.	Starling S. vulgaris indicus Blyth	WM	R
	Family: CORVIDAE		
69.	House Crow *Corvus splendens splendens* Vieillot	Rst	C
70.	Himalayan Jungle Crow *Corvus macrorhynchos intermedius* Adams	Rst, AM	C

Contd.

71.	Indian Jungle Crow C. *macrorhynchos culminatus* Sykes	Rst	C
72.	Northwestern Tree Pie *Dendrocitta vagabunda* (Blyth)	Rst	C
73.	Yellow Billed Blue Magpie *Cissa flavirostris* (Blyth)	Rst	C
	Family: CAMPEHAGIDAE		
74.	North Indian Scarlet Minivet *ericrocotus flammeus speciosus* (Latham)	Rst, AM	O
75.	Small Minivet *Pericrocotus cinnamomeus* (Linnaeus)	Rst	O
	Family: PYCNONOTIDAE		
76.	Red Vented Bulbul *Pycnonotus cafer cafer* (Linnaeus)	Rst	C
77.	White-cheeked Bulbul *P. leucogenys leucogenys* (Grey)	Rst	C
78.	Black Bulbul	Rst	O
	Hypsipetes madagascariensis ***(P.L.S. Muller)***		
	Family: MUSCICAPIDAE **Subfamily: TIMALIINAE**		
79.	Jungle Babbler *Turdoides striatus somervillei* (Sykes)	Rst	C
80.	Common Babbler *Turdoides caudatus caudatus* (Dumont)	Rst	C
	Subfamily: MONARCHINAE		
81.	Paradise Flycatcher *Terpsiphone paradisii paradisii* (Linnaeus	SM	R
	Subfamily: MUSCICAPINAE		
82.	Verditer Flycatcher *Muscicapa thalassina thalassina* Swainson	Rst, AM	R
	Subfamily: SYLVINAE		
83.	Indian Tailor Bird *Orthotomus sutorius guzuratus* (Latham)	Rst	O
	Subfamily: TURDINAE		
84.	Blue Rock Thrush *Monticola solitarius* (Linnaeus)	Rst	R

Contd.

85.	Himalayan whistling thrush *Myiophonus caeruleus* (Scopoli)	Rst, AM	C
86.	Brown Rock Chat *Cercomela fusca* (Blyth)	Rst	C
87.	Collared Bush Chat *Saxicola torquata indica* (Blyth)	WM	O
88.	Pied Bush Chat *Saxicola caprata bicolor* (Sykes)	Rst	C
89.	Indian Robin *Saxicoloides fulicata cambaiensis* (Latham)	Rst	C
90.	Indian Magpie Robin *Copsychus saularis saularis* (Linnaeus)	Rst	C
91.	White Capped Redstart *Chaimarrornis leucocephalus* (Vigors)	Rst, AM	C
92.	Plumbeous Redstart *Rhyacornis fuliginosus* (Vigors)	Rst	O
93.	Kashmir Black Redstart *Phoenicurus ochruros*	WM	C
94.	Western Spotted Forktail *Enicurus maculatus* (Vigors)	Rst	O
95.	Blue Throat *Erithacus = (Luscinia) svecicus svecicus* (Linnaeus)	WM	R
	Family: MOTACILLIDAE		
96.	Indian White Wagtail *Motacilla alba dukhunensis* Sykes	WM	C
97.	Indian Pied Wagtail *Motacilla madreruspatensis* Gmelin	WM	C
98.	Blue Headed Yellow Wagtail *Motacilla flava beema* (Sykes)	WM	R
99.	Grey Wagtail *Motacillla cinerea* Tunstall	WM	C
	Family: ZOSTEROPIDAE		
100.	Indian White Eye *Zosterops palpebrosa palpebrosa* (Temminck)	Rst	O
	Family: NECTARINIIDAE		
101.	Purple Sunbird *Nectarinia asiatica asiatica* (Latham)	Rst	C

Contd.

102.	Yellow-backed Sunbird *Aethopyga siparaja seheriae* (Tickell)	Rst, AM	O
	Family: PASSERIDAE **Subfamily: PASSERINAE**		
103.	Indian House Sparrow *Passer domesticus* (Linnaeus)	Rst	C
104.	Himalayan Cinnamon Tree Sparrow *P. rutilans* (Temminck)	Rst	O
	Subfamily: PLOCEINAE		
105.	Indian Baya *Ploceus philippinus* (Linnaeus)	Rst	C
	Subfamily: ESTRILDINAE		
106.	Spotted Munia Lonchura punctulata (Linnaeus)	Rst	C
107.	Red Munia Estrilda amandava amandava *(Linnaeus)* Family: PARIDAE Subfamily: PARINAE	*Rst*	*R*
108.	Grey Tit *Parus major* (Linnaeus)	Rst, AM	O
109.	Green Backed Tit *P. montlcolus* Vigors	Rst	R
	Family: FRINGILLIDAE **Subfamily: EMBERIZINAE**		
110.	Himalayan Rock Bunting *Emberiza* c/a Linnaeus	Rst, AM	O
111.	Crested Bunting *Melophus lathami* (Gray)	Rst, AM	R
	Family: HIRUDINIDAE		
112.	Red Rumped Swallow *Hirundo daurica erythropygia* Sykes	Rst	C
113.	Wire Tailed Swallow *Hirundo smithii* Stephens	Rst	O
	Family: CINCLIDAE		
114.	Himalayan Brown Dipper *Cinclus pallasii* Bonaparte	Rst	R

Contd.

	Family: ALAUDIDAE		
115.	Crested Lark *Galerida cristata* (Franklin)	Rst	C
	Family: CERTHIDAE		
116.	Himalayan Tree creeper	Rst, AM	R
	Order: BUCEROTIFORMES **Family: BUCEROTIDAE**		
117.	Common Grey Hornbill	Rst	R

Tockus (= Ocyceros) birostris (Scopoli)

Of the 117 species reported, 91 species were resident, recorded through out the year, 13 species show seasonal altitudinal migrations between low and high altitudes. Ten species have been reported to be winter migrants visiting the study area during winters only. Three species were reported to be summer migrants visiting the study area during summers only (Table 1).

As far as the local status of the birds reported is concerned, the study reveals that 55 species are Common, 31 species are Occasional and 31 species are Rare of the total species recorded.

Present work also reveals the range extension of 2 species of the reported "birds i.e. *Large Green Barbet* and *Yellow Backed Sunbird* as earlier records show their north limit extending upto Himachal Pradesh only. Large Green Barbet was recorded to be resident bird in the study area. Grimmett *et al.*[15] reported its distribution to start from Himachal Pradesh in the subcontinent. In the study area, it was reported from Hiranagar, Kathua, Basohli and Billawar within 240-650m range and was found in well wooded country, forests, gardens etc. in towns, cities, villages etc.

Similarly, Yellow Backed Sunbird is resident to the study area showing altitudinal migrations. Whistler[4], Ali and Ripley[6] and Grimmett *et al.*[15] do not mention J&K in the species distribution. During the present course of study, it was recorded from all the stations with an altitudinal range 350-2000m.

So, the above mentioned two species i.e. Large Green Barbet and Yellow Backed Sunbird show an extension in the distributional range.

The study area despite its small size appears to support an extremely rich and diverse bird community. Kathua district (2651 sq. kms) represents

0.008085% of the total geographic area of the country (3,27,87,263 sq. kms) however, avifauna of district Kathua represents 9% of the total birds (i.e. 1300 sp.) recorded from Indian subcontinent[15]. The observed bird diversity in relatively small study area underlies the importance of this area for biodiversity conservation.

Defining The Status

For assigning status to the faunal elements observed during the survey of study area based on sight records, the terminology of Khan (2002) was followed:

C - Common: means it can invariably be seen in that habitat where it occurs with the provision that the season is also appropriate.

F - Frequent: means that even visiting appropriate habitat, it will not be seen or heard invariably, perhaps only in one visit out of three.

O - Occasional: means seen or heard only in one visit out of six.

R - Rare: means even less likelihood of occurrence.

Besides this, depending upon whether the species of birds are sighted in all months / season of year or only in particular season / some months of the year and absent in others from the study area, it *SM - Summer Migrants:* Those which visit the study area in summers. *W/M - Winter Migrants:* Those which visit the study area in winters. *AM - Altitudinal Migrant:* Those which migrate between higher and lower altitudes.

Acknowledgement

We are thankful to the Department of Zoology, Univ. of Jammu, Jammu for providing various facilities to carry out the present study.

References

1. Ward, A. E. (1906a & 1906b). *J. Bomb. Nat. Hist. Soc.* 17: 108; 479.
2. Ward, A. E. (1907a & 1907b). *J. Bomb. Nat. Hist. Soc.* 17(3): 723; 17(4): 943.
3. Osmaston, B. B. (1927). *J. Bomb. Nat. Hist. Soc.* 31: 975.
4. Whistler, H. (1928). In: Popular Handbook of Indian Birds. London: Gurney and Jackson.

5. Ali, S. and Ripley, S.D. (1968-74). In: The Handbook of Birds of India and Pakistan. Ten Volumes. Oxford University Press, Bombay.
6. Ali, S. and Ripley, S.D. (1983). In: Compact Handbook of Birds of India and Pakistan. Oxford Univ. Press, Bombay.
7. Sahi, D.N. (1985). In: Breeding ecology of Blue Rock Pigeon, Columba Livia Gmelin. Jammu University Review, 3: 64.
8. Alfred, J.R.B, Kumar, A., Tak, P.C. and Sati, J.P. (2001). In: Water birds of Northern India. Zoological Survey of India.
9. Choudhary, V. (2002). In: Studies on the Avian Diversity of Jammu District of J&K State. Ph.D Thesis. University of Jammu, Jammu.
10. Sharma, B. (2003). In: Faunal Diversity of Ramnagar Wildlife Sanctuary, Jammu. M.Phil. Dissertation, University of Jammu, Jammu.
11. Ahmed, A. (2004). In : Diversity and Community Structure of the Birds of Tehsil Doda, Jammu. M.Phil. Dissertation. Univ. of Jammu, Jammu.
12. Wani, A.A. and Sahi, D.N. (2005). *J. Natcon.*, 173(1): 135.
13. Kumar, S. and Sahi, D.N. (2006). *J. Himalayan. Ecol. Sustain. Dev.*, 1: 95.
14. Ali, S. (1996). In : The Book of Indian Birds (12th and enlarged centenary edition). J. Bomb. Nat. Hist. Soc. Oxford Univ. Press, New Delhi.
15. Grimmett, R., Inskipp, C. and Inskipp, T. (1998). In: Birds of The Indian Subcontinent. Oxford Univ. Press, Delhi.
16. Grewal, B., Harvey, V. and Pfister, O. (2002). In: A Photographic guide to the Birds of India. Periplus Editions (HK) Ltd. Singapore.
17. Inskipp, T., Lindsey, N., and Duckworth, W. (1996). In: An Annotated Checklist of the Birds of the Oriental Region. Oriental Bird Club, Sandy.

17

Loss in Avian Biodiversity with Special Reference to Vanishing of Vultures in Eastern U.P. India

J. P. SHUKLA, A. K. MISHRA* & NAMRATA NATH TRIPATHI**

P. G. Department of Zoology; S.H. Kisan P. G. College, Basti (D.P.)
*P.G. Department of Zoology, L.B.S.S.P.G. College, Anand Nagar, Mahrajganj (U.P.) ** M.L.K.P.G College, Balrampur (U.P.)

Abstract

A close survey for last five years was made and found that the birds especially *Coracias bengalensis* (Nilkanth) of order Coraciformes, *Dinopium* (Woodpecker) of order Piciformes, *Ploceus phillipinus* (Baya) and *Passer domesticus*. (House sparrow) both of order Passeriforms are significantly declining. However, the Vultures (*Gyps bengalensis* order Falconiformes), famous as natural scavenger have almost vanished from Eastern U.P. Causes of loss and possible measures for the conservation of aforementioned avian species have been discussed.

Introduction

There are 2100 species of birds inhabiting India and many of them are severely endangered basically due to habitat loss, preying, amplification of agrochemicals like pesticides etc among avian species through food chain. Bamboo partridge *(Bambusicola futchi)* and the red spurfowl *(Galloperdix spadicea)* widely distributed[1] are fast turning rare. Keeping in view the loss in avian diversity, a survey was made for five years, 2001-2005. The avian species selected in our study were confined to *Ploceus phillipinus* (Baya); *Passer doemsticus* (House sparrow); *Coracias bengalensis* (Nilkanth) and *Dinopium* (Woodpecker). Special attention was given to monitor the existence of vultures *(Gyps bengalensis)* in the selected three districts of eastern UP. India.

Methods

Measurement of selected avian population was followed by the standard methods[2]. Since Basti happened to be once the largest area wise district of U.P. and now divided into three districts namely Santkabir Nagar, Siddhartha Nagar and Basti proper, our study was confined in all these three districts to record the abundance of afore mentioned avian species during the months of March/April when majority of the avian species for various purpose like feeding, nesting, mating etc. Generally take place. For each station of observation, an area of about 20 acres was taken in consideration. In Basti, Ban Bihar area, in SantKabir Nagar, Bakhira area and in Siddharthanagar, Rudhauli area or dense vegetations having variety or trees, was selected. Observations were made is between 9.00 am - 11.00 am and 3.30 p.m to 5.30 p.m. for five years (2001-2005). Mean values for 20 observations for each species, with S.D. are proved in Table 1. Ten days observation for four hours were conducted at each station.

Results and Discussion

Data given in Table-1 reveal that there is gradual decline in the population density of the selected regional avian species in the three districts in eastern *U.P.* (Basti, Siddharthanagar and Santkabirnagar). The vultures have entirely vanished. The four selected avian species may be grouped in the category of V (vulnerable) i.e, taxa which are likely to move into the endangered category in coming year, if the causative factors continue operating.

The selected avian species except vultures which are natural scavanger, are insectivorous in their feeding habits. They feed usually upon various insect pests of various crops and fruits. It is also an established fact that presently farmers use synthetic insecticides/agrochemicals instead of herbal insecticides/domestic or natural manures on one hand causing catastrophic alterations in the biosystems. On the other hand, due to rapid urbanization and construction etc., the natural habitat of various avian species is getting disturbed. Further, due to increase in human population, there develops hindrance in the natural ecosystem homeostasis. However, the most relevant factor responsible for the gradual decline in the selected avian population may be attributed to the frequent application of synthetic insecticides of various groups that represent biological amplification and get accumulated in the selected species in particular and on other biota in general.

Table 1. Population density of some bird species in three districts of Eastern U.P. during March-April, 2001-2005

Year	*Basti*			
	P. phillipinus	*P. domesticus*	*C. bengalensis*	*Dinopium sps.*
2001	42±24	68 ±4.2	24 ±3.2	20±2B
2002	36±2.0	52±4.0	22±3.4	16±2.6
2003	28±1.8	46±3.6	20±2.6	14±2.1
2004	26±2.1	40±3.2	18±2.2	11 ±1.8
2005	20±1.9	28±2.4	14±2.4	7±1.4
Year	*Siddharth-Nagar*			
	P. phillipinus	*P. domesticus*	*C. bengalensis*	*Dinopium sps.*
2001	56±4.2	75±4.4	26±2.6	16±2.0
2002	44±4.0	70±2.8	16±3.4	14±1.8
2003	38±3.1	62±3.4	14±24	8 ±2.4
2004	30±2.6	52±2.8	8±2.0	6±2.1
2005	22±4.0	38±3.4	6±1.2	4±1.2
Year	*Siddharth-Nagar*			
	P. phillipinus	*P. domesticus*	*C. bengalensis*	*Dinopium sps.*
2001	52±3.8	80±3.6	26±3.2	16±2 4
2002	46±24	72±2.4	18±2.0	13±2.1
2003	40i4.0	60±4.2	14 ±2.2	12±1.2
2004	26±44	44±3.6	10±1.8	10±1.2
2005	20±2.2	32±34	6±2.1	6±1.4

The amplified synthetic insecticides in the avian species may interfere with the hormonal synthesis causing impairment in the ovulation. Also these may cause fragility in the egg shell and hence the selected avian species have become vulnerable. It is also reported that after getting the entry of synthetic insecticides and certain agrochemicals through food chain in the target avian body, these chemicals interact with genetic material and cause mutation[3]. These substance not only affect the DNA and the chromosomes but represent biological amplification and get accumulated in the selected species in particular and on other biota in general.

The amplified synthetic insecticides in the avian species interfere with the hormonal synthesis causing impairment in the ovulation. Also these may cause fragility in the egg shell and hence the selected avian species are becoming vulnerable. Also after getting the entry of synthetic insecticides and certain agrochemicals through food chain in the target avian body, these may interact with genetic systems and may cause mutation These substances not only affect the DNA and the chromosomes but may have far reaching long-term influence on the progeny of the affected individuals.

Synthetic agrochemicals produce a reaction in the form of nucleophilic substitution of the first and / or second order (SN1, or SN2) by way of electrophilic cryptic ions in the genetic systems of the target animals which they attack[9].

Pesticides induce genetic changes due to their alkylating properties as these carry one or more alkylating groups. An alkylation- of the DNA in vitro by various alkylating agents normally occurs at N-7 atom of guanine this leads to increase in the acidity of N-7 position and appearance of ionized guanine, which thereupon becomes incapable of pairing with cytosine and ultimately pairs with thymine instead. It has also been established that alkylating agents react with the phosphoryl groups of the DNA and bring transformations in the replication kinetics[10-12]. Chromosomal pulverization also occurs due to intoxication of various chemicals causing impairment in the spermatogenesis and oogenesis in the target species. Decline in the selected four avian species may be largely due to the pesticides/agrochemicals consumed by these birds directly or indirectly through food chain and to certain extent due to the destruction of the habitat.

In our close survey no more seen in the study area vultures in fact, vultures are valued in Indian society for their specific role being regarded as natural scavengers. There are eight species of vultures, reported in India[19]. These are *Sacrogyps calvus, Aegypus monachus; Neophron percnopterus; Gyps fulvus; Gyps himalayensis; Gyps indicus; Gyps tenuirostris and Gyps bengalensis or Pseudogyps bengalenesis.* In eastern UP., *during the last decade of* earlier than the end of the past century, *Gyps bengalensis* were of common occurrance but at present, these are entirely absent from larger part of the state. The vultures feed on carcasses of cattle. After feeding they used to return on their shelters made on bamboo trees or Tarkul of Cycadaceae family and pearched silently in groups.

However, most of the eologists are of opinion that diclofenac, anti-inflammatory drug commonly used in eastern U.P. for treatment of livestock maybe the prime cause for vulture (*Gyps bengalensis*) decline. Veterinary diclofenace might have shown biological amplification in vultures, feeding upon carcass that would have interfered with the gametogenesis in male/female vultures. Urbanization and consequently habitat destruction may also be one of the causes for their decline.

Needless to mention, that other than diclofenac, many other deleterious toxicants of non - degradable nature have also got their way into the vultures body through food chain and cansed bioamplification, ultimately causing death of vultures due to various physiological disorders including impairment in gametogenesis, fragility of the egg shell etc. It advised to make very sincercere efforts for the conservation of remaining of the vultures in other parts of India. For eastern UP., authors wish to recommend the setting up of a centre to keep vultures in captivity for observation and experimentation in order to detect the disease and control of drugs as well as captive breeding. Phasing out of diclofenac and establishment of more vulture breeding centres, would hopefully ensure the survival of the vanished species.

Acknowledgements

Authors are thankful to Prof. K. Pandey, Prof. K. Kumar, Prof. C.P. M. Tripathi, Prof. Ajai Kumar Srivastava, Prof. R. Singh, Dr. D.K. Singh, Dr. Ajai Singh of Zoology Department, Deen Dayal Updhayay Gorakhpur University, Gorakhpur for their suggestions to investigate the avian biodiversity at regional level.

Reference

1. Sunder, I. (2004). Env. Pollution. Tech. 3, 353-358. In everyman's Science. Vol-XLI No. 5, p-322-327 by G. Tripathi (2006-2007).
2. Kendeigh, S. Chartes (1944). Ecol Monogr. 14:67.
3. Sharma, A.K. (1981) In: Proceeding: Impact of development of science and technology on the environment 68th session of ISCA held at Varanasi, (India).
4. Barker, C.J. & Rackam, B.D. (1979). Mut at. Res., 68:381.
5. Hsu, T.C. (1982). In: Cylogenetic assays of environmental mutagens. Oxford and L.B.H. Pub. Co. New Delhi.

6. Prasad, A.B. (1986). In: Mutagenesis- Basic and Applied. Print House India, Lucknow.
7. Sahai, R. (1990). Pollut. & Health Hazard. III.
8. Gupta, S.C. & Sahai, R. & Gupta, N. (1996). Aus. Asian J. Animal. Sci., 9(4): 449.
9. Vogel. F. & Rohrborn, G. (1970). In: Chemical mutagenesis in mammals and man (eds) Heidelberg Springer, New York (USA).
10. Elmore, D.T., Gulland, J.M., Jordn D.O., & Taylor, H.W.F. (1948). Biochem. J., 42.
11. Ross, W.C.J. (1953). In: Adv. Cancer Res., 1 : 397.
12. Reiner, B. & Zamenhof, S. (1957): *J. Biol. Chem.*, 228 : 475
13. Gupta, S.C. & Jain, N. (1983). Mechanisans of pesticide induced mutations. In Proa, 8th annual Conf. of Env. Mutagen Soc. of India, Hyderabad.
14. Shukla, J.P. & Pandey, K. (1984a). Toxicol. Lett. (London) 21:1.
15. Shukla, J.P. & Pandey, K. (1984b). Toxicol. Lett. (London) 21 : 191.
16. Shukla, J.P. & Pandey, K. (1984c). Cellular & Molecular Biology. 30: 227
17. Shukla, J.P. & Pandey, K. (1984d). Bull. Inst. Zool. Acad. Sinice. 23:69
18. Shukla, J.P. & Pandey, K. (1986). Acta Hydrochim et. Hydrobiol 14:195
19. Chhangani, A.K., Rajpurohit LS. & Mohnot, S.M. (2005). Ind. Sci. Cong., p. 27 Ab. No. 39

18

Diversity of Medicinal Flora of District Pithoragarh & Their Uses By The Local Communities

KAMAL KISHOR GANGWAR AND B.D. JOSHI
Department of Zoology and Environmental Sciences, Faculty of Life Science, Gurukula Kangri University Haridwar, UK

Abstract

A total of 69 species of plants of medicinal importance, belonging to 38 families were identified during surveys made around fifteen locations in district Pithoragarh. It was found that *Rubus ellipticus, Rhododendron arboreum, Solanum indicum, Asparagus racemosus, Pyracantha crenulata, Withania somnifera, Ocimum kilimandscharicum, Micromeria bioflora, Thymus surphylum, Aloe vera* are being used widely to cure various diseases by the local community in the area under study.

Introduction

The Indian Himalayan region alone supports about 18,4,40 species of plants. (8000 species of Angiosperms, 44 species of Gymnosperms, 600 species of Pteridophytes, 1736 species of Bryophytes, 1159 species of Lichens and 6900 species of Fungi) of which about 45% are having medicinal values. According to Samant *et al.*[1] out of the total species of vascular plants, 1748 species are medicinal. Many medicinal plants species are rapidly losing ground against the onslaught of man's exploitation and encroachment. There also exists a thriving market for medicinal plants. To meet ever growing demand, medicinal plants are usually collected from the wild stock. Records indicate that as much as 95% raw materials required by Pharmaceuticals and drug manufactures are collected from the wild[2]. Most of these are collected using non sustainable, destructive methods like collecting the entire plant, rhizome, timber, roots or other reproductive parts like fruits and seeds. Such destructive activities are the major factors influencing floral populations. Further, low regeneration rate and loss of habitat add to serious threat to the existence and distribution of medicinal plants. Such as rapid depletion of medicinal plants from the natural habitat required urgent conservative measures. Studies related to conservation of medicinal plants have been carried out by many workers[3-6]. A comprehensive review[7] has described the rich diversity and use of medicinal flora with in Uttarakhand, besides a recent study on the medicinal flora in the riparian zone of river Ganga at Saptrishi (Haridwar) by Gangwar and Joshi[8].

Himalayan Herbal Medicines and Their Use by Local Peoples

Traditional Himalayan medicines have not only helped and saved the lives of poor people in these far flunged mountainous rural areas but also of rich peoples around the world. About 80% population of the world depends on the traditional health care system. World Health Organization has also estimated that approximately 80% of the people in developing countries depend on traditional medicines for primary health care needs[7-9]. Allopathic drugs have brought a revolution around the world but the plant based medicines have their own status. Traditional knowledge of Himalayan medicine is a good illustration of poor communities, fighting even incurable diseases through the traditional methods, and even for their livestock, through these traditional herbal medicines.The indigenous traditional knowledge of medicinal plants of various local communities has been transmitted orally for centuries is becoming extinct, almost globally, due to introduction of modern technology and changes in the traditional culture. The use of plants for treatment in India dates back to prehistoric times. This indigenous knowledge about medicinal plants and therapies was composed verbally and passed orally from generation to generation.

Concept of Traditional Himalayan Herbal Medicines

Diseases are the bane of humankind ever since its advent on this planet. Man has been fighting against a variety of diseases since prehistoric periods. Eventually humans developed an indigenous system of folk medicines.

In certain areas these folk medical prescriptions are purely indigenous and have survived through ages from one generation to the next through the word of mouth. Most of these are not available in the form of written knowledge. Generally these systems of medicine depend on old people's experiences. Indigenous systems of medicine are specially conditioned by the cultural heritage and myths. Much later, some of this information was systematized in the form of treatise like Atharveda, Yajurveda, Charak Samhita, Sushrut Samhita, etc. These systematized systems of knowledge about medicinal plants and therapies are included under Ayurveda - the Indian Traditional Medicine System. Despite the development of rural health services, villagers still use herbal native medicines to a large extent for treatment of common ailments like cough, cold and fever, headache and body-ache, constipation, dysentery, burns, cuts and scalds, boils and ulcers, skin diseases and respiratory troubles, etc.

With a deep concern and reverence for the vast plant diversity that our country enjoys and with sense of realization about the invaluable therapeutic properties of this plant diversity, the current research has been conducted under a special assistance program of UGC for last three years. This work concentrates on the distribution of plants of medicinal importance commonly used by the local communities in and around district Pithoragarh, situated at the border line of Indo-China-Tibbat and Nepal in the state of Uttarakhand.

Methodology

Pithoragarh district is situated at an altitude of 1650 meters above sea level. The field visits were made during the year of 2006-2007. The surveys for the medicinal plants were conducted at fifteen selected spots viz. Dharamgarh, Dhamara, Kanalicheena, Ogla, Jauljeevi, Baluakot, Dharchula, Tapovan, Tawaghat, Chirgala, Sovla and around Swaminarayan Temple at Chhota Kailash, Ainchioli, village Dhamora and village Dhari of district Pithoragarh. The information regarding the tradition of medicinal plants available in the local area for treating various diseases was collected by directly contacting the elders, herbalists and the local peoples who have knowledge about these medicinal plants. Identification of plants was done by local experts and with the help of available literature[5,10-16].The medicinal value of each plant was enumerated in the following pattern: (*i*) Botanical name, (*ii*) Family, (*iii*) Vernacular name/local name, (*iv*) Plant parts used ans (*v*) Medicinal uses.

Table 1. Ethno-medicinal plants used by the local communities of District Pithoragarh

S. No.	*Family & Botanical Name*	*Vernacular/ Local Name*	*Plant Parts Used*	*Therapeutic Uses*
		Fam. 1. Agavaceae		
1.	*Agave americana*	Rambans	Leaves	Fever and Boil
		Fam. 2. Apiaceae		
2.	*Centella asiatica*	Mandukparni	Whole plant parts	Skin diseases and to improve memory
		Fam. 3. Apocynaceae		
3.	*Holarrhena antidysenterica.*	Kura/Dudhi	Root Stem, Bark and seed	Diarrhea, dysentery, haemorrhoid, rheumatic arthritis, skin diseased.
		Fam. 4. Araceae		
4.	*Acorus calamus* Linn.	Buch	Root	Commercial
5.	Arisaema tortuosum (Wallich) Schott	Bag-mugri	Whole Plant	Fever, paste of tubers applied on burns
6.	*Arnebia benthamii* (Wallich Ex)	Balsamjari	Root	Hair tonic
		Fam. 5. Asteraceae		
7.	*Artemisia nilagirica* (Clarke) Pamp.	Nagdona	Leaves, flower & whole plant	Hysteria, epilepsy, nervous irritability and nervous depression.
8.	*Artimisia capillaries* Thunb.	Sitabani	Whole plant	Purgative and smoke of plant used on burns.
9.	*Cichorium intybus* (Linn)	Kasni	Seed, roots and flowers	Digestion, eyesight, decoction of the powdered seeds is used in obstructed or disordered menstruation. The flower syrup is given in liver disorders.
10.	*Erigeron asteroids Roxb.*	Bangua	Seeds & roots	Used in stimulant for diuretic.
11.	*Eupatorium odoratum*	Tivra gandha	Oil of plant	To control of worms.

	Fam. 6. Berberidaceae		
12. *Berberis asiatica* (D.C)	Rasanjana/ Daruhaldi/ Kilmora	Root bark, Stem, wood and fruits	Malaria, indigestion, skin disease, leucorrhoea, Jaundice, severe, snake bite, Manly vesedmeye, Ear, oraleasity diseases.
	Fam. 7. Compositae		
13. *Sonchus olereaceous* L.	Dudhi	Leaves and stem	Lever diseases
	Fam. 8. Dioscoreaceae		
14. *Dioscorea deltoidea* Wall.	Ban tarur	Tuber	Urinogenetal disorders
	Fam. 9. Dipsacaceae		
15. *Morina polyphylla* L.	Kandru	Roots	Used in wounds
	Fam. 10. Ephedraceae		
16. *Ephedra gerardiana* Wall, ex Stapf	Tut gatha	Stem	Asthma
	Fam. 11. Ericaceae		
17. *Rhododendron arboreum (S.M.)*	Buransh /Brash	Flowers Juice	Stomach Diseases
	Fam. 12. Euphorbiaceae		
18. *Euphorbia royleana* Boiss.	Sulu	Latex	Antiseptic, germicidal, stop bleeding, ear complaints, hollow cavities of tooth
19. *Ricinus communis* L.	Arand	Root, Root bark, leaves, flowers, fruits, seeds & oil	Leaves used: disuria, cough, worm infestation. Fruit used: hepatomegaly, spleenomegaly, epilepsy, piles, Asthma, bronchitis, skin diseases
	Fam. 13. Gentianaceae		
20. *Swertia chireta* Roxb. Syn.	Chireta	Dried whole plant, roots, stems and flowers	Malarial fever, leprosy, leucoderma, scabies, neuromuscular, monorrhagia, menstrual irregularity, urinary disease, heart disease, indigestion, asthma, cough, dyserasia, ulcer, jaundice, anemia.

		Fam. 14. Guttiferae		
21.	*Hypricum podocarpoides* N. Robson	Tikua	Whole plant	Used to control worms
		Fam. 15. Hippocastanacaeae		
22.	*Aesculus indica*	Pangar	Roots	Roots used for leucorrhoea and rheumatism
		Fam. 16. Lamiaceae		
23.	*Colebrookia oppositifolia* Smith	Binda	Leaves and roots	Leaves applied to worms and bruises, roots used in prescription for epilepsy
24.	*Mentha longofolia Linn.*	Wild pudina	Whole plant, Stem and Leaves	Antifertility, Anti-ovulatory, gastrointestinal disorders, dyspepsia, cough, cold, chronic fever
25.	*Micromeria biflora* Benth	Van ajwain,	Leaves	Worm infested, cold & sinusitis
26.	*Ocimum kilimand-scharicum* Guerke.	Kapoor tulsi	Oil	Pains and sprains
27.	*Ocimum sanctum*	Ram tulsi	Leaves and twings	Used in diarrhea, astringent, rheumatism
28.	*Salvia lavata*	Panya	Root	Rheumatism
29.	*Thymus surphylum* Linn.	Wild thyme	Leaves and floral shoots	Whooping cough and epilepsy
		Fam. 17. Liliaceae		
30.	*Asparagus filicinus* Buch.-Ham. Ex D. Don	Sharanoi	Roots	Diarrhoea, dysentery and diabetes
31.	*Asparagus recemosus* (Willd.)	Satawar/ Sahasmuli	Flashy roots and Cladodes	Threatened abortion, leucorrhoea, seminal debility, General debility, hysteria, Useful in acidity and ulser patient extract of cladode is Anticancer.
		Fam. 18. Malvaceae		
32.	*Malvastrum coromandilianum Garcke.*	Bala	Leaves and stem	Dysentry, wounds, jaundice, sprain, and rheumatism
		Fam. 19. Moraceae		
33.	*Ficus carica*	Anjir	Fruits	To improve health
34.	*Ficus religiosa*	Pipal	Fruits	To bear children for unfertile women

		Fam. 20. Myricaceae		
35.	*Myrica esculanta* (Buch Ham.)	Kaphal	Bark	Chronic cough, asthma, painful dental gin and ear ache, external application in healing of chronicg and malignant ulcers.
		Fam. 21 . Orchidaceae		
36.	*Dactylorhiza hatagirea* (D. Don) Soo	Hathajari	Roots	Burns/cuts
		Fam. 22. Papilionaceae		
37.	*Desmodium elegans* DC.	Sambar	Root	Epilepsy
		Fam. 23. Parnassiaceae		
38.	*Parnassia nubicola* Wallich ex Royle	Nirbis .	Root	Antidote for poison
		24. Pinaceae		
39.	*Pinus roxburgii. Roxb*	Chir/Sarala	Sargent and Resin	Stomach ache, effective on kidney, polyurie, arthritis, colic pain, swellings, abscess, Boils, wound healing, ulcers
		Fam. 25. Poaceae		
40.	*Eliliopsis binata*	Bhabhar ghas	Roots	Fever and liver complains
		Fam. 26. Polygonaceae		
41.	*Rheum australe* D. Don	Chhirchey	Root	Wound
42.	*Rumex histatus* D. Don	Chalmori	Leaves	To cure eye diseases
43.	*Rumex nepalensis* Spr.	Khatura	Root	Purgative
		Fam. 27. Punicaceae		
44.	*Punica granatam (Linn.)*	Anar	Juice, fruit bark & flower	Fruit juice used in piles, bark and flowers in dysentery, decoction of flower buds used in bronchititis and vaginal discharges

		Fam. 28. Ranunculaceae		
45.	*Aconitum heterophyllum* Wallich Ex Royle	Atis	Root	Headache
46.	*Paeonia emodi* Wall	Hilto	Root	Uterine disease
47.	*Ranunculus sceleratus L.*	Jaldhania	Leaves	Rheumatism, dysuria, asthma and pneumonia
48.	*Thalictrum foliolosum* DC.	Mamira	Root	Purgative
		Fam. 29. Rosaceae		
49.	*Rubus ellipticus*	Hisal/Hisalu	Roots	Blood pressure, diarrhoea
50.	*Rosa sericea*	Wild rose	Flowers	For skin care
51.	*Prensepia utilis Royle.*	Jhatalu	Whole plant	Oil used in Rheumatism and in paindue to fatigue
52.	*Pyrus malus*	Seb	Fruits	Edible
53.	*Prunus prasica*	Andu	Leaves	Wormicidal in humans
54.	*Cotoneaster microphyllus Wall Ex. Lindn*	Wanni	Leaves extract, Fruits	Diarrhoea, strigent
		Fam. 30. Rutaceae		
55.	*Zanthoxylum armatum* DC.	Timur	Fruit	Toothache
		Fam. 31. Sapindaceae		
56.	*Sapindus mukorossi Garten.*	Reetha	Seeds	Hair and skin diseases
		Fam. 32. Saxiferagaceae		
57.	*Berginia strachyi* (Hookf.& Thoms.) Engl.	Pashanbheda	Rhizome and bark	Diuretic, opthelmia, cuts and antiscorbotic
		Fam. 33. Scrophulariaceae		
58.	*Bocopa monieri* (Linn.)	Brahammi/ Jal neem	Whole plant	Dyspepsia, Cough fever, Insomnia, epilepsy, debility after heart attack, hoarsness of voice, less memory tension, blood purifier.

59. *Picororrhiza kurroa* Royle ex Benth.	Kutki	Roots & Rhizome	Dyspepsia, asthma, jaundice, heart disease, malarial fever and worms infestation in children
60. *Verbascum thapsus* Linn.	Ekalveer	Whole Herb	Asthma and pulmonary diseases
	Fam. 34. Solanaceae		
61. *Hyoscyamus niger* L.	Langtang	Seed	Toothache
62. *Solarium indicum* Linn Syn.S.ferox Linn.	Bhata katari	Fruits	Asthma, Dry cough, colic, disuria, chronic fever, alopecia and dropsy
63. *Solanum nigrum* Linn.	Makoi	Fruits	Dysentery and fever
64. *Withania somnifera* Dunal.	Ashwagandh a	Roots	To improve memory in humans
	Fam. 35. Umbrlliferae		
65. *Pleurospermum angelicoides*	Chhipi	Root	Stomach ache
	Fam. 36. Urticaceae		
66. *Urtica dioica*	Bichchhua	Leaves	Uterine hemorrhages, bleeding from nose and blood vomiting
	Fam. 37. Valerianaceae		
67. *Nardostachys grandiflora DC.*	Masi	Leaves	Stomach ache
	Fam. 38. Zingiberaceae		
68. *Hedychium spicatium* Ham ex Smith.	Gandh Palasi/ Kapoor Kachri	Rhizome	Hypoglycemia, inflammation, cough, asthma, vomiting, colic, ulcer, hysteria, dental trouble,
69. *Roscoea alpina* Royle	Kakoli	Roots	Rheumatism

Results and Discussion

The data on medicinal plants which were collected from the fifteen selected locations within the Pithoragarh district were pooled and their enumerations along with ethno-medicinal uses are described in Table-1. The investigation revealed the local people, herbalists and vaidyas have explored a number of plant species to curing various ailments. As a result of present survey, a total of 69 plant species of ethno-medicinal importance, belonging to 38 families are documented from above-mentioned twelve spots of district Pithoragarh of Uttarakhand. The primitive knowledge about medicinal plants is so vast that more than 3000 plant species in whole of India have been listed as useful for population control[18]. During a preliminary survey of Bhagirathi valley in the Himalayas, 61 medicinal plants belonging to 37 families were reported by Uniyal[19]. Chhetri[20] reported 281 species of ethno-medicinal plants belonging to 108 families are used in folk medicines in the Darjeeling Himalaya. Thus it is very apparent that a treasure of herbal medicines is spread in these mountain valleys of our country especially in areas of states like Uttarakhand.

Acknowledgement

This work has been carried out under the UGC-SAP programme, sanctioned to the Department of Zoology and Environmental Sciences,Gurukula Kangri University Haridwar for which we are thankful to University Grants Commission, New Delhi for providing financial support during the period of study.

References

1. Samant, S.S., U. Dhar & L.M.S. Palni. (1998). In : Medicinal Plants of Indian Himalaya: Diversity, Distribution Potential Value. Himavikas Publication No. 13. Gyanodaya Prakashan, Nainital.
2. Kehimker, I. (2000). In : Common Indian Wild flowers, Bombay Nature History Society, Oxford University Press, Walton Street, Oxford, OXZ26DP.
3. Dhiman, A. K. (1997). In: PhD thesis, *A survey of medicinal plants of Haridwar and adjoing area vis a vis the raw plants drugs being sold in local market.* Gurukul Kangri University, Haridwar.
4. Brahmverchas. (2003). In: *Ayurved ka Pran: Vanausdhi vigyan,* Shantikunj, Haridwar.
5. Dhiman, A.K. (2005). In: Wild medicinal plants of India(With ethnomedicinal uses). Published by Bishan Pal Singh and Mahendra Pal Singh, 23-A, New Connaught Place Dehradun.
6. Singh, Y.P. and Joshi, B.D. (2005). *Him. J. Env. Zool.*, 19(1): 97.
7. Joshi, B.D. (2002). *Him. J. Env. Zool.*, 16(2): 233.
8. Gangwar, R.S. and Joshi, B.D. (2006). *Him. J. Env. Zool.*, 16 (2):233.
9. Raven, P. H. (1992).The nature and value of biodiversity. 1-5 in global biodiversity strategy. World Resource Institute. WCU/UNEP.

10. Purohit, S.S. and Vyas, S.P. (2005). In: Medicinal plants and cultivation. Agrobios India Jodhpur.

11. Polunin, O and Stainton, A. (2005). In: flowers of the Himalaya (7th Ed.). Oxford University Press, New Delhi.

12. Kanjilal, U.N. (2004). In: Forest flora of the Chakrata, Dehradun & Saharanpur forest division (Foreword by Pradeep Krishen). Natraj Publisher, Dehradun.

13. Nair, C.K.N. and Mohanan, N. (1998). In: *Medicinal plants of India*, Nag Publishers, Delhi.

14. Rajiv, K. (1988). In: Economic Plants in Vedas. Common wealth Publisher, New Delhi.

15. Faulks, P.J. (1958). In: An introduction to Ethnobotany. Moredale Publication London.

16. Jain, S.K. (1995). In: A manual of Ethanobotany (Second edition). Scientific Publishers, Jodhpur.

17. Sandhya, B., Thiomas, S., Isabel, W. and Shenbagarathai (2006). *Afr. J. Trad. CAM.*, 3 (1): 101.

18. Rahmani, A.R. (1970). *Tiger Paper.* 79 (2):24.

19. Uniyal, M.R. (1968). *Indian Forester.* 94: 407.

20. Chhetri, D.R. (2005). Current Science. 89 (2): 264.

19
Species Richness and Abundance of Earthworms in Cultivated Soils in Kumaun Himalayas

RENU BISHT, D. BHARTI, H. PANDEY, S. BORA AND B.R. KAUSHAL
Department of Zoology, D.S.B. Campus, Kumaun University, Nainital-263002

Abstract

Earthworm population dynamics and their activity in cultivated soil of the Kumaun Himalaya were studied. Ten species of earthworms were found: two lumbricids, seven megascolecids, and one moniligastrid. The maximum density (212.8 m^{-2}) and biomass (71.6 g m^{-2}) of earthworms were recorded in the wet season. An average of 96.0% of total worm density and 96.2% of worm biomass occurred in 0-10 cm soil at Simalsar, white 92.8% of density and 90.9% of biomass occurred in 0-10 cm at Kungaon. Significant correlation were observed between earthworm density and biomass with soil temperature, and soil organic content in the present investigation.

Introduction

Earthworms are an important component of soil fauna being well known for ameliorating soil fertility by enhancing the physical, chemical and biological characteristics of soil and for their action in soil turn-over by depositing casts on the soil, nutrient cycling and in plant litter decomposition[1]. Earthworms can comprise a significant portion of the total biomass (80-96%) of invertebrates in some soils[2]. Most published studies on earthworm ecology are from temperate regions[3] Few studies have been reported from tropical soils in India[4,5] and cultivated soils of Kumaun Himalayas[6]. This paper deals with: (1) earthworm assemblages and population dynamics, (2) vertical distribution of earthworms in the soil, (3) relations between earthworm density and biomass with pedological characteristics for developing management strategies for improving soil fertility in Kumaun Himalayas.

Materials and Methods

Earthworms were sampled at Simalsar (29°23' N, 79°31' E, and altitude 1400 m) and at Kungaon (29°20' N, 79°35' E, altitude 1560 m). Two crops are grown each year at the study sites: paddy (late June-early October) and wheat and seasonal vegetables (November - April). The tillage system in both sites is manual and the tilled depth is about 5 cm.

Earthworms were collected from 5 random collections fortnightly by hand sorting using a quadrat of 50 × 50cm, from depths of 0-10 and 10-20 cm. The earthworms were rinsed, dried on blotting paper, weighed (including gut content) and then preserved in 4% formaline[7]. The collected specimens were identified following[5]. All individuals < 3 cm without a clitellum were categorised as aclitellates. Climatic data were recorded at the nearest meteorological station, UP. State Observatory, Nainital. On the basis of climatic variations, the year is divisible in to three seasons, summer (March-June), rainy (July-October) andwinter (November-February). The average annual temperature and precipitation are 14.8 ^{0}C and 1769 mm, respectively. The soil is alfisol of laterite origin at both sites and contains 51.6% gravel, 35.9% sand, 9.5% silt and 3.0 % clay at Simalsar; 58.4% gravel, 28.2% sand, 9.8% silt and 3.6% clay at Kungaon. Soil samples were taken from 5 random collections from the two sites and depths (0-10 and 10-20 cm) at each earthworm sampling date every fortnight. Soil temperature, soil moisture and pH were measured on each sample date. Soil was air-dried, ground, passed through a 2 mm sieve and stored for subsequent analysis. pH was determined on a 1 : 25 soil water suspension by pH meter. Soil moisture was determined gravimetrically drying at 105 ^{0}C. Air-dried, ground soil (passed through a 2 mm sieve) was used for chemical analysis. Organic C was determined using wet oxidation, P using the wet-ashing method or and soil N was determined by Kjel auto Vs-KTP Nitrogen Analyser based on micro-Kjeldahl technique[9]. K was determined by flame photometry. Regression (r) between worm density and biomass and soil properties were derived from correlation coefficient.

Results and Discussion

Pedological Characteristics

Soil pH was near neutral in both soil layers in both sites (Table 1). Maximum soil moisture content was 35.3% at 0-10 cm and 33.9% at 10-20 cm soil depth of Kungaon; and 35.9% at 0-10 cm and 25.9% at 10-20 cm soil depth at Simalsar.Organic C decreased with increasing depth. C:N ratio increased with increasing depth at Simalsar but showed a decline at Kungaon. It is clear from Table 1 that pH, temperature and moisture do not show significant differences within or between sites, but K, P, organic C and N show significant differences.

Species Composition and Abundance of Earthworms

A total of 883 earthworms from 165 samples at Simalsar and 1927 worms form 170 samples were collected at Kungaon, 10 species were identified (Table 2) belonging to three families *(Lumbricidae-2* species, *Megascolecidae-7* species and *Moniligastridae*-1 species) in the two sites. Percent composition of *Lumbricidae* was 5.2%, *Megascolecidae* 60.4%, and aclitellates 34.4% at Simalsar; and *Megascolecidae* 49.2%, *Monligastridae* 2.6%, and aclitellates 48.2% at Kungaon. The abundance and biomass of earthworms at 0 - 20 cm depth are given in Figs. 1 and 2. At Simalsar, worm density ranged from 0 m^{-2} (12 July) to 107.2 m^{-2} (24 September) (Fig.1) and from 0 m^{-2} (31 May to 212.8 m^{-2}) at Kungaon (Fig. 2). The maximum density of worms were recorded in the wet period and minimum density in the winter and summer seasons at both sites. It is presumed that earthworms would be present throughout the year, but when soils are dry and or temperatures are high earthworms would have retreated to depth > 20 cm. However, sampling for earthworms at > 20 cm soil depth was not undertaken in the present study.

Earthworm biomass ranged from 0 g m^{-2} (12 July) to 62.7 g m^{-2} (10 October) at Simalsar (Fig. 1) and from 0 g m^{-2} (31 May) to 71.6 g m^{-2} (3 September) at Kungaon (Fig. 2). Thus, in both years, the maximum density and biomass of earthworms were recorded during or towards the end of the wet period and minimum values were obtained during dry period (winter and summer seasons) at both sites. The seasonal dynamics over an annual cycle showed that the earthworm population and biomass were high in the wet period and low in summer and winter. The present study showed a preference of earthworms to soil sown to paddy rather than to wheat crop. Higher abundance of earthworms under paddy may be related to higher reproduction and survival due to higher moisture there.Typical dimensions suggested for the surface area of soil to be removed are 50 × 50 cm[1]. The depth of the soil removed should match the vertical distribution of the earthworm species. Sachel[1] pointed out that in hand-sorting, small earthworms (<2 cm in length) are more often overlooked than larger ones. Hand sorting can be supplemented by combining it with formalin application. Formalin application repels higher earthworm numbers and biomass, especially individuals of *Lumbricus terrestris* in rich soils, than the hand-sorting method[10]. A combination of these two methods was used in the present study, the efficiency of which was tested by[11]. This method has the advantage that it can be adapted to extent earthworms at all stages of the life cycle. It provided the population in the upper 20 cm of soils in the present study.

Vertical Distribution of Earthworms in the Soil

Vertical distribution of earthworms in the two crops and of both crops combined together are given in Table 3. A mean of 95.7% of the worms and 96.0% of the biomass were recorded in the 0-10 cm soil layer in the paddy crop at Simalsar. A higher proportion in number (99.2%) and biomass (98.7%) of worms was recorded in the 0-10 cm soil layer in the wheat and seasonal vegetables crop. A mean of 96.0% of total worms and 96.2% of the total biomass occurred at a depth 0 -10 cm in comparison with numbers and biomass individuals 10-20 cm. At Kungaon, a mean 97.1% of total earthworms and 96.4% of the total biomass were recorded in the 0-10 cm soil layer in the paddy crop. In the wheat and seasonal vegetables crops, a mean of 74.6% of total earthworms and 74.2% of the total biomass were recorded in the 0-10 cm soil layer. A mean of 92.8% of total earthworms and 90.9% of the total biomass of earthworms were recorded at a depth of 0-10 cm in comparison with number of individuals at 10-20 cm (Table 3). Higher density and biomass at the 0-10 cm layer in comparison with the number of individuals at the 10-20 cm layer in both sites indicates that inhabitants of 0-10 cm layer had a relatively large body size. The habitat preference of lumbricid species was studied in various localities in Europe are several groupings of these species have been proposed[12] but system of classifying earthworms[13] into three categories: epigees - litter dwellers, endogees - horizontal burrowers and anecique - deep burrowers is most widely used.The habitat preference of earthworms was clear in the studied sites in the present investigation. The earthworm species are mineral-soil living since 73-100% of the specimens were recorded at 0-10 cm in depth. The proportion of biomass in the 0 - 10 cm layer was also higher (70-100%) than in the 10-20 cm soil layer. Most researchers have found earthworms almost exclusively in the top 50 cm of agricultural soils, and most species in the top 20 cm[22] The data obtained in the present work fit this pattern for the earthworm population as a whole. Difference in the vertical distribution of the C:N ratio at 0-10 cm and 10-20 cm, and the organic matter seems to be most important factor in earthworm distribution.

Age Structure

Only two age classes have been considered, aclitellates and clitellates. At Simalsar, the yearly ratio of clitellates to aclitellates in the 0-20 cm soil in the paddy crop was 1: 0.51, and 1: 0.65 in the wheat and seasonal vegetables crop. At Kungaon, the yearly ratio was 1: 1.1 in the paddy crop, and 1: 0.27 in the wheat and seasonal vegetables crop. However, when two crops were considered together, the yearly ratio was 1: 0.53 at Simalsar and 1: 0.93 at Kungaon (Table 4). As a rule, in annual cycles of the ratio of aclitellates, the number of adults is below 50% of the overall count for any species. This was the case only in the paddy crop at Kungaon whereas at Simalsar (both in the paddy crop and wheat and seasonal vegetable crop), adults constitute higher proportion of the total worms. It was further observed that numbers of clitellates declined in winter and summer seasons (dry periods). In case of aclitellates, the winter drop in number was rapid and abrupt in the wet period, the number of aclitellates increased rapidly.

Table 1. Some soil characteristics in the cultivated soils (n = 34; mean ± S.E.).

Soil Characteristics	*Simalsar*		*Kungaon*	
	Soil layer		*Soil layer*	
	0-10 cm	*10-20 cm*	*0-10 cm*	*10-20 cm*
Soil pH	6.8 ±0.1	6.7 ± 0.2	7.3 ±0.1	7.3 ±0.08
K (%)	0.04 ±0.003	0.028 ±0.002	0.050 ±0.01	0.028 ± 0.003
P (%)	0.0052 ±0.001	0.0057 ±0.001	0.00518 ±0.001	0.0081 ±0.002
Organic C (%)	3.05 ±0.1	2.6 ±0.1	4.2 ±0.1	3.3 ±0.2
N (%)	0.40 ±0.02	0.3 ±0.03	0.44 ±0.02	0.38 ± 0.013
C : N ratio	7.6	8.7	10.0	8.7

Table 2. Number of individuals of different species of earthworm collected in cultivated field soils.

Species	*Simalsar*		*Kungaon*	
	No.	%	No.	%
Lumbricidae				
Aporrectodea rosea Savigny	37	4.2	-	-
Octolasion tyretaeum Savigny	9	1.0	-	-
Megascolecidae				
Amynthas corticis Kinsberg	43	4.9	209	10.8
Amynthas gracilis Kinsberg	98	1 1.1	-	-
Amynthas morrisi Beddard	-	-	71	3.7
Metaphire anomala Mich.	201	22.7	71	3.7
Metaphire birmanica Rose	119	13.5	168	8.7
Metaphire houletti Perr.	59	6.7	387	20.1
Perionyx excavatus Perr.	13	1.5	43	2.2
Moniligastridae				
Drawida japonica Mich.	-	-	50	2.6
Aclitellates	304	34.4	928	48.2
Total earthworms collected	883	100.0	1927	100.0

Table 3. Vertical distribution of earthworm density (individuals m^{-2}) and biomass (gm^{-2}) in the cultivated soils of two localities. Figures in parentheses are percentage values.

Year	*Sfinalsar*				*Kunga on*			
	0-10 cm		*10-20 cm*		*0-1D cm*		*10-20 cm*	
	Density	*Biomass*	*Density*	*Bi. amass*	*Density*	*Biomass*	*Density*	*Biomass*
Paddy crop	305	97.29	10	1.8	576	168.66	0	0
1996-97	453	161.42	26	10,65	9 64	171.65	60	13.54
1997-98	(95.7)	(96.0)	(4.3)	(4.0)	(97.1)	(96.4)	(2.9)	(3.6)
Mean								
Wheat and seasonal vegetables	28	1 1.71	0	0	97	40.76	36	17.2
	60	12.94	1	0.34	148	48.21	46	12.89
1996-97	(99.2)	(98.7)	(0.8)	(1.3)	(74.6)	(74.2)	(2 5.4)	(25.8)
1997-98								
Mean								
Total worms of both crops	333	109.0	10	1.8	673	209.42	36	17.2
	513	174 36	27	10.99	1 1 12	220.16	106	26.43
1996-97	(96.0)	(96.2)	(4.0)	(3.8)	(92.8)	(90.9)	(7.2)	(9.1)
1997-98								
Mean								

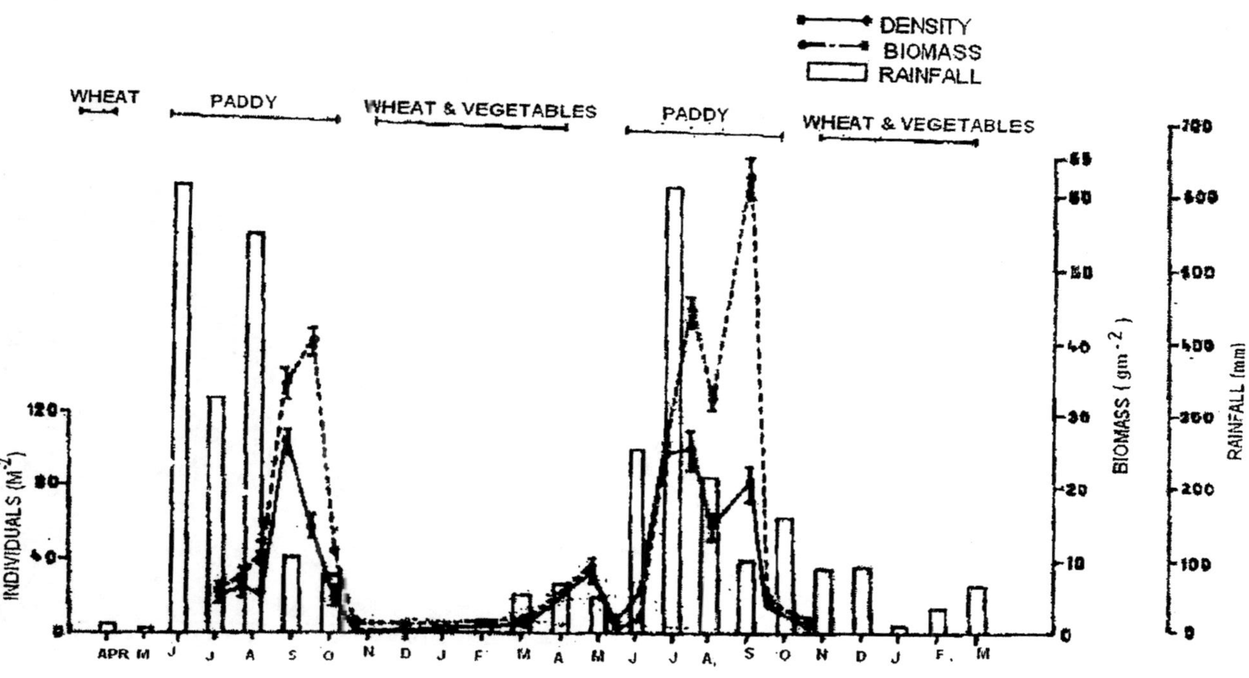

Fig. 1: Earthworms (Ind.m^{-2}), fresh biomass (g m^{-2}) and rainfall (mm) in 0-20 cm in cultivated soils at Simalsar.

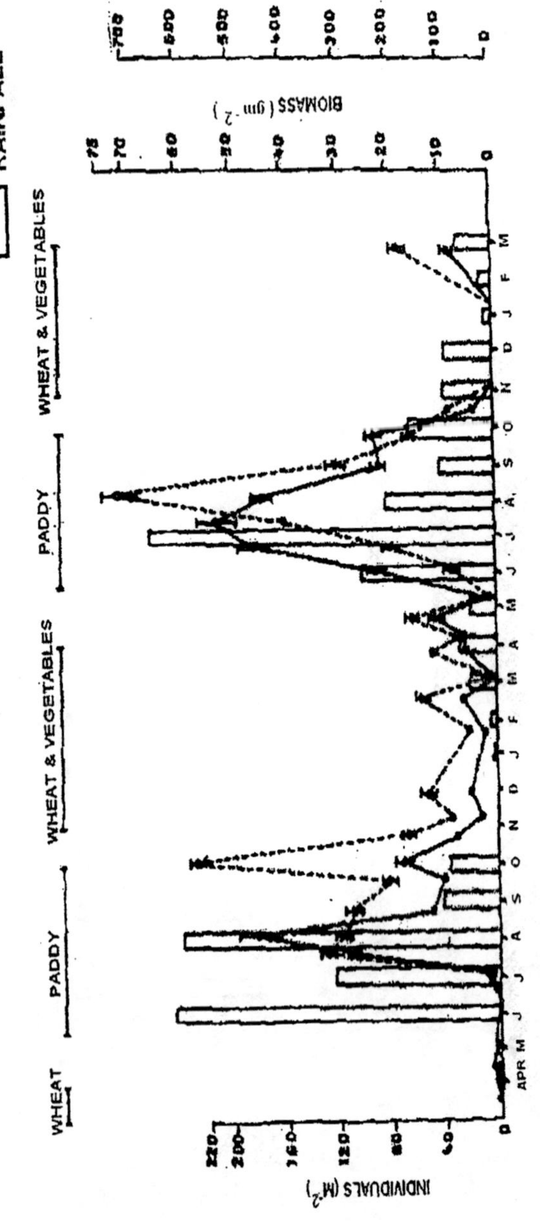

Fig. 2: Earthworms density (Ind. m-2), fresh biomass (g m-2) and rainfall (mm) in 0-20 cm in cultivated soils at K[illegible]ngaon.

Table 4. Age structure of earthworms in the cultivated soils. Figures in parentheses are sample size

Sites	*Total worms collected (O-2O cm)*		*Clitellates*	*Aclitdlates*
Siinalsar				
Paddy crop	794	(60)	525	269
Wheat and seasonal vegetables	89	(105)	54	35
Total of two crops	883	(165)	579	304
Kuugaon				
Paddy crop	1600	(65)	327	(105)
Wheat and seasonal vegetables	759	240	841	87
Total of two crops	1927	(1T0)	999	928

Juveniles generally predominate in the population of epigeic and anecic earthworm species, e.g. *Lumbricus terrestris*[3,14]. In the present study, mature individuals were dominant throughout the year and immature only during summer and rainy period. This characteristic is typical for species within the endogeic oligohumic category[15].

Relations Between Earthworm Density, Biomass and Soil Characteristics

Correlation analysis were performed between earthworm density and biomass and soil characteristics of the 0-10 cm soil layer because maximum density and biomass of earthworms were recorded in this layer.

Correlation analysis indicated that worm density was significantly negatively correlated with soil C ($n = 12, r = - 0.571; p < 0.05$); and worm biomass significantly negatively correlated with soil C ($n = 12, r = - 0.716; p < 0.05$) in the paddy crop at Simalsar. At Kungaon, significant positive correlations between worm density and soil temperature ($n = 34, r = 0.506; p < 0.01$); significant negative correlations between worm density and soil C ($n = 34, r = -0.451; p < 0.01$) and worm biomass and soil C ($n = 34, r = - 0.443; p < 0.01$) when data on worm density and biomass of earthworms of the two crops were combined together. A variety of environment such as soil texture, soil moisture, food, pH, temperature, soil depth, organic content have been suggested as determinants of the distribution and abundance of worms[6-18] Significant positive correlations were also observed between earthworm density and biomass with soil temperature and negative correlations with organic content in the present investigation also.

Acknowledgement

The authors express their thanks to the Department of Science and Technology, New Delhi for financial assistance through a research project to Dr. B.R. Kaushal, and to Dr. J.M. Julka for identifying earthworm species.

References

1. Satchell, J.E. (1971). In: Soil Biology. (Ed. A. Burges & F. Raw). Academic Press, London.
2. Didden, W.A.M., Marinissen, J.C.Y., Vreeken Buijs, M.J., Burgers, S. L.G. E., Fluiter, R. and Brussard, L. (1994). *Agriculture, Ecosystem & Environment* 51: 171.
3. Edwards, C.A. (1983). Chapman and Hall, London.
4. Dash, M.C. and Patra, U.C. (1977). *Revue a" Ecologie et Biologie du Sols* 14 : 461.
5. Julka, J.M. (1988). Zoological Survey of India, Calcutta, India.
6. Kaushal, B.R., Kandpal, B., Bisht, S.P.S., Bora. S. and Dhapola, R. (1999;. *European Journal of Soil Biology.* 35 : 171.
7. Rozen, A. (1982). *Paedobiologia* 23 : 199.
8. Edwards, C.A. and Lofty, J.R. (1982). *Soil Biol. & Biochem.* 14: 515.
9. Jackson, M.L. (1958). Soil Chemical Analysis. Prantice Hall, Englewood Clifts, N.J., U.S.A.
10. Raw, F. (1959). *Nature* (London). 184 : 1661.
11. Raw, F. (1960). *Nature* (London). 187 : 257
12. Julin, E. (1948). *Acta Phytogeography Succ.* 23 : 186.
13. Bouche, M. (1972). *National des Recherches Agricultturelles,* Paris.
14. Daniel, O. (1992). *Soil Biology & Biochemistry.* 24 : 1425.
15. Lee, K.E. (1985). Earthworms: their ecology and relationship and land use. Academic Press, Sydney, Australia.
16. Blanchart, E. and Julka, J.M. (1997). *Soil Biol. & Biochem.* 29: 303.
17. Lavelle, P. (1983). In: *Earthworm Ecology* (Ed. J.E. Satchell). Chapman and Hall, London.
18. Mele, P.M. and Carter, M.R. (1999). *Applied Soil Ecology.* 12 : 129.
19. Misra, R. (1968). Ecology Workbook. Oxford and IBH Publishing Company, Calcutta, India.

20

Diversity of Medico-Economically Important Riparian Floral of River Ganga at Shyampur, Dehradun

RADHEY SHYAM GANGWAR AND B. D. JOSHI

Department of Zoology and Environmental Science, Faculty of Life Sciences, Gurukul Kangri University, Haridwar-249 404, Uttarakhand

Abstract

The present study records riparian floral diversity along the river Ganga, which has economic and medical value for humans. A total number of 97 floral species belonging to 47 families and 85 genera at the both banks of river Ganga were recorded and identified. The maximum numbers of species are documented from family Malvaceae (6 sp.) followed by Euphorbiaceae (5 sp.), Asteraceae (5 sp.) and Solanaceae (4 sp.).

Introduction

The riparian region has become the place for most frequent human activities. Plants that grow along the river banks are called riparian vegetation[1]. Riparian plant diversity plays a significant role to minimizing the water pollution related to water body to check solid waste and chemicals (fertilizers & pesticides), which come from adjacent upland through run off. The floral composition of a forest constitutes the most important aspect of all the biotic composition of a healthy ecosystem[2]. Information on the economic aspects of plants has been passed from one generation to the next generation without any published records[3]. Since ancient period people depend on plants for most of their basic needs. It is universal fact that no living organism can survive without plants. Human beings have been continuously using the plants for various requirements. The study of plants in the service of mankind has been a part of human civilization. A study was made on riparian medicinal floras in Haridwar[4]. Studies on medicinal plant diversity were conducted by some earlier workers[5-7]. This study was carried out in the riparian zone of Ganga near Shyampur village, which comes under the Dehradun district of Uttarakhand state in Garhwal Himalaya and located at the right bank of the river Ganga at an elevation of 315 meter from sea level. The position of village on globe is situated 30°, 03' N and 78° 14' E. In the present study an attempt has been made to determine riparian floral diversity along Ganga River, which has medicinal and economic values.

Materials and Methods

Sample of each species were collected and identified with the help of local experts and the available literature[8-10]. A through literature survey was made to compile existing information on the medicinal and economic uses. In addition, local persons also acted as a source for recording ethnobotanical information of each plant for various purposes. Seasonal visits to the above mentioned place were made from January 2005 to December 2006. Ethnobotanical value of each plant species has been enumerated in the following pattern: (*a*) Botanical Name, (*b*) Family, (*c*) Vernacular Name, (*d*) Ethnobotanical uses.

Results and Discussion

During the course of present investigation, a total of 97 plant species belonging to 47 families and 85 genera were documented. The maximum numbers of species are recorded from family Malvaceae followed by Euphorbiaceae, Asteraceae and Solanaceae. All identified species of plants have been categories for different purposes such as medicinal and economic importance. Out of 97 plants species, most of them are used for multi-purposes. The riparian plant diversity of river Ganga varies at place to place from the distribution point of view. In tree canopy, some patches were dominated by *Acacia catechu,* while few patches were dominant with *Bombex ceiba* and *Dalbergia sissoo.* Among the shrubs layer, some spots were dominant with *Lantana camera, Adhaiods zeylanica* and *Tamarix dioica* and *Cassia occidentalis.* It also was observed during the survey that grazing of cattle and fuel collection by local people are threats to riparian vegetation. The inventory and utilization of these plants are furnished in Table 1 below.

Table 1. Riparian floral diversity and their uses for medicinal and economic values.

S. No.	*Botanical Name*	*Family*	*Vernacular Name*	*Ethnobotanical uses*
1.	*Abutilon indicum* (L) Sw.	Malvaceae	Kanghi	Source of fiber and is considered as laxative
2.	*Abrus precatorius* Linn.	Papilionaceae	Ratti	Seeds are used for beads.
3.	*Acacia catechu* (L.f.) Willd.	Mimosacee	Khair	It is used for tanning, toilet preparations, dyeing fabrics and also used in medicine.
4.	*Achyranthes aspera* L	Amaranthaceae	Latgira	Roots are used to relieve pain from scorpion stings and pounded roots in water are given in stomach pain.
5.	*Adhatoda zeylanica* Medic	Acanthaceae	Adusa	It is used as a medicine for bronchitis, asthma and fever
6.	*Adina cordifolia* Hook, f.	Rubiaceae	Haldu	It is used for construction, furniture and agricultural implements
7.	*Aegle marmelos* (L.) Corr.	Rutaceae	Bel	Ripe fruits used in sherbets and drinks. Leaves also used as a fodder for cattle.
8.	*Aerva lanata*	Amaranthaceae		Shoots are used in the treatment of piles given with pepper and milk.
9.	*Ageratum conyzoides* L.	Asteraceae	Visadodi	It is used for cuts and sores.
10.	*Alternanthera* DC. *sissilis* (L.)	Amaranthaceae	Gudrisag	It is used as pot herb.
11.	*Amaranthus virdis* L	Amaranthaceae	Chaulai	The shoots and leaves are consumed as pot-herb
12.	*Anisomeles indica* (L.) 0. Kuntz.	Lamiaceae		Medicinal
13.	*Anogeissus latifolia* Wall.	Combretaceae	Bakali	It is used extensively for construction, furniture, agricultural implements and leaves also used as a fodder for cattle.
14.	*Apluda mutica* L.	Poaceae	Charol	Veterinary mouth sores.

15.	*Ardisia solanacea* Roxb.	Myrsinaceae		Roots used in diarrhea and rheumatism.
16.	*Argemone maxicana* L.	Papaveraceae	Katila	The oil of plant is used for skin diseases
17.	*Azadirachta indica* A. Juss.	Meliaceae	Neem	Seeds are source of margosa oil for skin diseases. It also used for soap making and leaves are inject repellant.
18.	*Boerhavia diffusa* L	Nyctaginaceae	Punarnava	Roots and leaves used as a medicine for peritonitis asthma and jaundice. Plants juice is good in eye troubles.
19.	*Bombex ceiba* L.	Bombacaceae	Semel	Wood used for paper making and roots in medicine for impotency
20.	Butea *monosperma* (Lamk) Taub.	Papilionaceae	Dhak	Leaves are used for making plates in villages and bark is used in snake bites. Seeds are laxative and anthelmintic.
21.	*Callicarpa macrophylla* Vahl	Verbenaceae	Daia	Fruits edible
22.	*Calotropis procera* (Air) R. Br.	Asclepiadaceae	Mada	Roots and bark used in dysentery
23.	*Cannabis sativa* Linn.	Cannabinaceae	Bhang	Leaves is used as tonic, intoxicant, in leprosy, diarrhoea and pains
24.	*Carissa opaca* Stapf.	Apocynaceae	Karaunda	Used for making combs. Fruits are used for pickles.
25.	*Casearia tomentosa* Roxb.	Flacourtiaceae	Chilla	It is used for furniture.
26.	*Cassia fistula* L.	Caesalpinaceae	Amaltas	It is an ornamental avenue tree and pulp of fruit used as purgative
27.	*Cassia occidentalis* L.	Caesalpinaceae	Kasaundi	Seeds used as a substitute for coffee
28.	*Cassia tora* L.	Caesalpinaceae	Chakunda	Used for skin diseases and seeds & leaves paste in water applied on eczema.
29.	*Chenopodium ambrosioides*	Chenopodiaceae		Used as a vegetable.
30.	*Chrysopogon fulvus*	Poaceae	Bhuri	Cattle Fodder grass

31.	*Colebrookia oppositifolia* Sm	Lamiaceae	Binda	The decoction of the root is used in epilepsy and leaves are applied on wounds and sores.
32.	*Cordia dichotoma* Forst. f.	Boraginaceae	Lasoora	The fruits are edible as vegetable and good in colic.
33.	*Crateva magna* (Lour.) DC.	Capparaceae	Barna	It is used for medicinal purpose
34.	*Curculigo orchioides* Gaertn.	Hypoxidaceae	Kalimusli	Rhizome is used for jaundice, piles and asthma.
35.	*Dalbergia sissoo* Roxb.	Fabaceae	Shisham	Used for carpentry work for high class furniture and cabinet. It is used for making charcoal.
36.	*Deshmostachya bipinnata* (Linn.) Stapf.	Poaceae	Kush	It is religious plant for Hindus and as a fodder for cattle.
37.	*Eclipta prostrata* (L.)	Asteraceae	Bhangraj	Decoction of the leaves in water is spread over the head of new-born children to stimulate for the growth of hair.
38.	*Ehretia leavis* Roxb.	Boraginaceae	Chamror	Wood used for match boxes, building & agricultural implements. Leaves are used as a fodder for cattle.
39.	*Emblica officinalis* Gaertn.	Euphorbiaceae	Aonla	Fruits are of much economic value, usually liteed for pickles.
40.	*Euphorbia hirta* L.	Euphorbiaceae	Dudhi	Used in medicine for cough and asthma.
41.	*Ficus benghalensis* Linn.	Moraceae	Bargad	Bark is used in ayurvedic medicines and milky latex is used in diarrhoea.
42.	*Ficus religiosa* Linn.	Moraceae	Pipal	It is sacred tree for Hindus and Buddhist. Bark and leaves used as astringent, rheumatism, purgative and in gonorrhea.
43.	*Glycosmis mauritiana* (Lamk.) Tanaka	Rutaceae	Ban-nimbu	Leaves are given for the treatment of fever and skin diseases
44.	*Helicteres isora L.*	Sterculiaceae	Kapasi	Fruits decoction is usted in traditional medicine for fever, cold and cough.

45.	*Holarrhena antidysenterica* Wall.	Apocynaceae	Kura	It is used for turning and furniture and seeds are used for dysentery.
46.	*Holoptelia integrifolia* (Roxb.) Planch.	Ulmaceae	Kanju	Wood for building purpose, carving, cabinet work, carts and combs
47.	*Hyptis sauveolense* (L.) Poit.	Limiaceae	Vilayti tulsi	It is used as green manure and taken as a beverage.
48.	*Imperata cylindrica*	Poaceae	Sirua	Used in medicine for dysentery.
49.	*Ipomoea carnea ssp. fistulosa* Mart, ex Choisy	Convolvulaceae	Sada Suhagan	Leaves externally as poultice in swellings.
50.	*Ipomoea nil*	Convolvulaceae	Kala dana	Seeds are used in traditional medicine.
51.	*Jatropa curcas* Linn	Euphorbiaceae	Safed Arand	In manufacturing of candles, soap and also in wood industry
52.	*Lagerstroemia parviflora* Roxb.	Lythraceae	Dhaudi	It is used in the Carpentry.
53.	*Lannea coromendelica* (Houtt) Merr.	Anacardiaceae	Jhingan	Decoction of bark used in body swellings
54.	*Lantana camera* L.	Verbenaceae	Lantana	Used for fuel and source of fiber.
55.	*Leea aspera L.*	Leeaceae	Kunwai	Ripe fruits are eaten.
56.	*Mallotus philipinensis* (Lam.) Muell.	Euphorbiaceae	Rohini	The bark is sometimes used for tanning. The crimson powder, kamela, which covers the ripe fruit, is used for dyeing silk.
57.	*Malvastrum coromandelianum* (L.) Garcke	Malvaceae	Bala	Used in medicine for dysentery and stem is source of fiber.
58.	*Martynia annua* L	Myrtyniaceae	Bichu	Fruits used in weaving block design in the baskets
59.	*Mitragyna parvifolia* (Roxb.) Korth	Rubiaceae	Phaldu, Kaim	Wood is used in preparation of agricultural tools.
60.	*Murraya koenigii* (L.) Spr.	Rutaceae	Kurry patha	Leaves and roots are used as tonic.

61.	*Nyctanthes arbor-tristis* L.	Oleaceae	Harsingar	The wood is used only for medicine.
62.	*Opuntia dilleni* Haw	Cactaceae	Nagphana	Fruits edible and also used in medicine
63.	*Oxalis corniculata* L	Oxalidaceae	Chuka	It used as pot herb and source of vitamin c.
64.	*Parthenium hysterophorus* L.	Asteraceae	Gazar Ghas	It is weed plant but used in medicine for dysentery.
65.	*Physalis minima* L.	Solanaceae	Tulatipati	Fruits consumed as vegetable
66.	*Pogostemon plecranthoides* Desf.	Lamiaceae	Ban tulsi	Leaves have the smell of black currants. The ashes prepared from the stems are used in some places as % manure in paddy nursery beds.
67.	*Polygon um hydropiper* L.	Polygonaceae	Water pepper	The leaves are diuretic used in uterine disorders and fish poisoning
68.	*Pupalia lappacea* L. Juss.	Amaranthaceae	Antreetha	Roots and fruits are used for rheumatic pains, inflammation, cough, fever and leprosy.
69.	*Pterospermum* acerifb//um Willd.	Sterculiaceae	Kanakchampa	The leaves are lopped for cattle fodder.
70.	*Saccharum munja*	Poaceae	Moonj	It is used for shed making.
71.	*Saccharum spontaneum* L.	Poaceae	Kans	Leaves employed for thatching brooms, mats
72.	*Sapium sebiferum* Roxb.	Euphorbiaceae	Tarcharvi	The leaves are used in black dye.
73.	*Scheichera oleosa* (Lour.) Oken	Sapindaceae	Kusum	Seeds are source of kusum oil and used as illuminant for manufacturing of soap and leaves of plant used as a fodder for cattle.
74.	*Shorea robusta* Gaertn. f.	Dipterocarpaceae		Sat Wood used for bridge construction,. boats, sleepers, furniture and other carpentry work.
75.	*Sida acuta* Burm. f.	Malvaceae	Karcta	It is source of fiber and also used in medicine for poultice of sores.

76.	*Sida cordata* (Burm. f.)	Malvaceae	Mahadaya	Leaves are used for healing cuts and juice with cow milk is taken as cooling & nervous tonic.
77.	*Sida rhombifolia* L.	Malvaceae	Kharenti	Leaves used as a tea and stem are source of fiber.
78.	*Solanum indicum* L.	Solanaceae	Barhanta	It is used as medicine in curries.
79.	*Solanum nigrum* L.	Solanaceae	Makoi	Its fruits are used for dysentery.
80.	*Solanum surratense* Burm. f.	Solanaceae	Kantakari	It is used in ayurvedic medicines as Dasamula for chest pain, cough, asthma and fever.
81.	*Sygyzium cumini* (Linn.) Skee	Myrtaceae	Jamun	Fruits are edible and seeds used as medicine as well as fodder.
82.	*Tamarix dioica* Roxb.	Tamaricaceae	Jhau	Wood used for fuel and branches used for making baskets.
83.	*Tectona grandis* Linn.	Verbenaceae	Sagaun	Wood durable and used for furniture, cabinetwork and construction
84.	*Terminalia alata* Heyne ex Roth	Combretaceae	Sain	Wood used for building purposes, agricultural implements, rail road sleepers, cabinet and other carpentry works.
85.	*Terminalia bellerica* (Gaertn.) Roxb	Combretaceae	Bahera	Fruits edible and leaves are used for tanning.
86.	*Trewia nudiflora* Linn.	Euphorbiaceae	Gutel	Wood used for manufacturing of drums, carved images and agricultural implements.
87.	*Tridax procumbens* L.	Asteraceae	Akal kohadi	Leaves used in dysentery and diarrhoea.
88.	*Triumfetta rhomboidea* Jacq.	Tiliaceae	Chiki	Bark and leaves used for diarrhoea and roots for dysentery.
89.	*Typha elephantine* Roxb.	Typhaceae	Patera	Leaves are used for mats making and hairs of fruits used for stuffing mattresses and pillows.

90.	*Urena lobata* Linn.	Malvaceae	Vilaytisan	Stem used for sacking, cordage, ropes and fishing tackle.
91.	*Vetiveraia zizaniodes* (L.)	Poaceae	Khas	It is used to make aromatic scented, mats, fans. ornamental baskets and Hindus used its in worship.
92.	*Vitex negundo* L.	Verbenaceae	Nirgundi	Branches used for manufacturing of baskets and also used in medicine.
93.	*Wrightia tomentosa* Roem. & Sen.	Apocynaceae	Dudhi	The bark of the stem and the roots are given as an antitode to snake bite and the sting of scorpions.
94.	*Xanthium strumarium* L.	Asteraceae	Kutta Ghas	Leaves are used in malaria and Urinary diseases
95.	*Zizyphus nummalaria* (Burm.f.)W. &A.	Rhamnaceae	Jharber	Leaves are used in malaria and Urinary diseases.
96.	*Zizyphus mauritiana* Lam.	Rhamnaceae	Ber	Fruits are edible and leaves used for tanning
97.	*Zizyphus xylopyra* Willd.	Rhamnaceae	Bhander	The bark is used for tanning and leaves for fodder.

Acknowledgement

This work has been carried out under the UGC-SAP programme, sanctioned to the Department of Zoology and Environmental Sciences, Gurukula Kangri University Haridwar for which we are thankful to University Grants Commission, New Delhi for providing financial support during the period of study.

References

1. Guh, Y. R., Zhao, W.C., Liang, S.H. and Shieh, C.L. (2005). *Int. J. Ecol. Environ. Sci.* 31 (1): 29-38.
2. Jòshi, B.D., Sharma, S.D., Joshi, K.D., Sengupta, S., Deorani, M.K. and Joshi, P.C. (1988). *Him. J. Env. Zool.* 2: 146-148.
3. Nayar, M.P., Ramamurthy, K. and Agarwal, V.S. (1989). Botanical Survey of India, Calcutta.
4. Gangwar, R.S. and Joshi, B.D. (2006). *Him. J. Env. Zool.* 20 (2): 237-241.

5. Sandhya, B., Thiomas, S., Isabel, W. and Shenbagarathai (2006). *Afr. J. Trad.CAM,* 3 (1): 101-114.
6. Singh, N.K. and Singh, D.P. (2001). *Flora and Fauna.* 7 (2): 59-66.
7. Singh, Y.P. and Joshi, B.D. (2005). *Him. J. Env. Zool.* 19 (1): 97-104.
8. Sahni, K.C. (2000). BNHS, Oxford University Press, New Delhi, 2nd edn.
9. Kanjilal, U.N. (2004). Natraj Publishers, Dehradun.
10. Kehimkar, I. (2000). BNHS, Oxford University Press.

21
Traditional Uses of Medicinal Plants in Almora District of Kumaun Himalaya, Uttarakhand

VIJAY SHARMA AND B.D. JOSHI
Department of Zoology and Environmental Sciences, Faculty of Life Sciences, Gurukula Kangri University, Haridwar (Uttarakhand), India.

Abstract

A medico botanical survey was carried out in the forest areas of Almora District in Uttarakhand State. The local people of study area were consulted to get information about the uses of medicinal plants. The present study reveals that 37 species of medicinal plants belonging to 30 families are being used traditionally for curing common ailments like asthma, diarrhoea, dental problems, urinary disorders, cold & cough etc. by the local community.

Introduction

Ancient civilization from all over the world and specially in the Indian system of medicine have been using herbal as well as animal and earth products as raw materials. The medicinal plants are recognized for their therapeutic value and immense potential in curing various ailments with almost no side effects. Medicinal plants have curative properties due to the presence of various complex chemical substances of different composition, which are found as secondary plant metabolites in one or more parts of these plants. Alkaloids are the most important chemical substances present in plants. Approximately 80% of the world's population rely on the traditional system of plant based medicines for their primary healthcare[1].

The scope of herbal medicinal utilization and its commercial aspects have reviewed by Joshi[2] quite recently. Almora district lies in the Uttarakhand State of Central Himalaya, which is well known for its natural vegetation particularly its medicinal plants. It lies between latitude 29° 32' 45" - 29° 53' 58" N and Longitude 79° 14' 25" - 79° 56' 7" E.

Materials and Methods

An Ethno medicinal survey was carried out during August 2006 to December 2006 in the forest areas of Kausani, Someshwar and Panwanaula regions of Almora district.

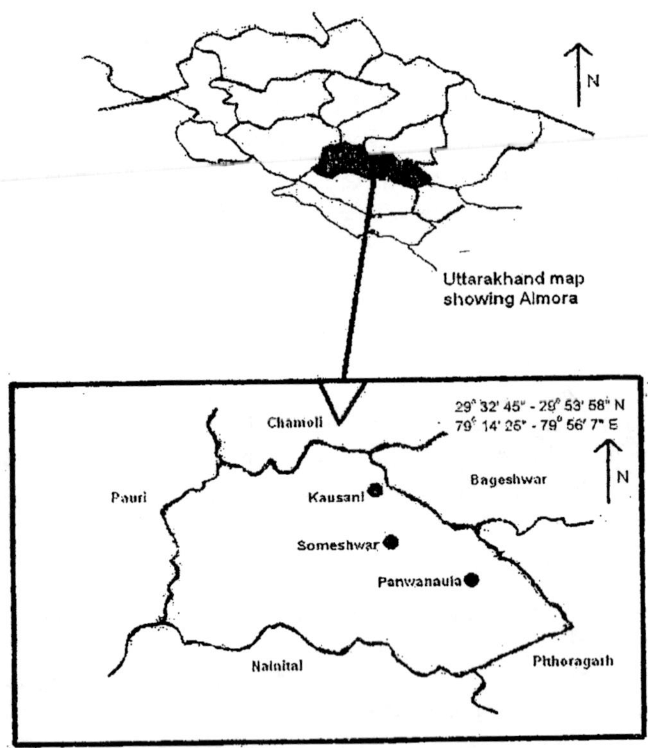

Location map of study area

The information regarding to the usage of medicinal plants for curing various ailments, was collected by directly contacting the elders and herbal healers in the study area. Voucher specimens were collected during field trips. The medicinal value of each plant was enumerated in the following patterns:

1. Botanical Name
2. Family
3. Local/Common Name
4. Ethno-medicinal Uses.

The identification of plants was done using the available literature[1,3-5].

Results and Discussion

The present study deals with the information of 37 plant species belonging to 30 families, which are used traditionally in Almora District for curing various ailments like Dental Problems, Asthma, Dysentery & Diarrhoea, Urinary disorders and Cold & Cough etc. The enumeration and utilization of these plants are described in Table 1.

Table 1. Enumeration and utilization of plants.

S. No.	*Botanical Name*	*Family*	*Local/Common Name*	*Ethnomedicinal Uses*
1.	*Acorus calamus* Linn.	Araceae	Vacha	Paste of root powder is given to the children for improving mental faculties. The rhizome is chewed to cure cough and sore throat.
2.	*Albizia lebbek* Benth.	Mimosaceae	Siris	Bark decoction is given in dysentery. Seed paste is usable in skin diseases.
3.	*Asparagus racemosus* Willd.	Liliaceae	Satawari	A special preparation made from the tubers is used to correct seminal weakness and impotency.
4.	*Berberis asiatica* Roxb ex DC.	Berberidaceae	Kilmora	Root extract is useful in eye diseases.
5.	*Bergenia ligulata* Engl.	Saxifragaceae	Pashanbheda	Rhizome extract is useful in urinary calculi.
6.	*Bombax ceiba* Linn.	Bombacaceae	Semal	Gum is useful in diarrhoea and dysentery.
7.	*Cedrus deodara* (Roxb.) Loud.	Pinaceae	Devdar	Tar is given in chronic skin diseases.
8.	*Centella asiatica* (Linn.) Urban.	Apiaceae	Mandukparni	The leaves or entire herb is boiled in water and its decoction is used as brain tonic.

9.	*Cinnamomum tamala* Nees & Ebesm.	Lauraceae	Tejpat	Leaf infusion is given in diarrhoea and bark infusion in cold & cough.
10.	*Cissampelos pareira* Linn.	Menispermaceae		Pari/Padi Roots are used in asthma and cold & cough.
11.	Citrus *medica* Linn.	Rutaceae	Nimbu	Powder of dried rind part of fruit is mixed with salt. This mixture is used as tooth powder in case of pyorrhea.
12.	*Cynodon dactylon* (Linn.) Pers.	Poaceae	Dubra	Fresh juice of grass stops bleeding from fresh cuts.
13.	*Datura stramonium* Linn.	Solanaceae	Dhatura	Seeds paste is applied to the infected teeth. Seeds mixed with butter are burnt and smoke is inhaled into the mouth in case of toothache.
14.	*Eclipta prostrata* Roxb.	Asteraceae	Bhringraj	Juice of crushed arieal parts (2 tea spoon full) mix with honey (2 tea spoon full) and given orally to cure cough.
15.	*Eucalyptus globules* Labill.	Myrtaceae	Safeda	Oil of leaves is antiseptic and used in ointment for burns.
16.	*Hedychium spicatum* Buch. Ham. ex Smith	Zingiberaceae	Van Haldi	Rhizome is useful in cough and asthma. It is also useful in headache.
17.	*Juglens regia* Linn.	Juglancaceae	Akhoda	Oil & fruits are used in making traditional tooth powder to cure toothache and pyorrhea.
18.	*Justicia adhatoda* Linn.	Acanthaceae	Vasing	Leaves and roots are useful in cough and asthma.
19.	*Myrica esculenta* Buch. Ham.	Myricaceae	Kaphal	Bark decoction is used in asthma and bronchitis.
20.	*Ocimum sanctum* Linn.	Lamiaceae	Tulsi	Infusion of leaves is applied to skin in ringworm and other cutaneous diseases.
21.	*Phyllanthus emblica* Linn.	Euphorbiaceae	Amla	Dried fruits are used in diarrhoea and dysentery.
22.	*Pinus roxburghii* Sarg.	Pinaceae	Chir	Oil is used as insencticide.
23.	*Prunus cerasoides* D.Don.	Rosaceae	Paiya	Stem bark is useful in stomach trouble and cold & cough.

24.	*Punica granatum* Linn.	Punicaceae	Anar	Edible portion of fruit is useful in brain affection. It is also useful in stop bleeding from nose.
25.	*Pyracantha crenulata* D.Don.	Rosaceae	Ghingaroo	Fruits are useful in cardiac failure and hypertension.
26.	*Quercus leucotrichophora* Cam.	Fagaceae	Banj	Leaves are useful in diarrhoea.
27.	*Rhododendron arboreum* Sm.	Ericaceae	Burans	Flower extract with water is usable in diarrhoea and dysentery.
28.	*Ricinus communis* Linn.	Euphorbiaceae	Arand	Oil is used in joint pain with swelling.
29.	*Rubus ellipticus* Sm.	Rosaceae	Hisalu	It is useful as antitoxic agent.
30.	*Solanum nigrum* Linn.	Solanaceae	Makoi	Ripe fruit is given orally to child in cough. Berries are useful in urinary disorders.
31.	*Solanum surattense* Burn. f.	Solanaceae	Kantkari	The plant is useful in bronchitis and urinary disorders.
32.	*Swertia chirata* Buch. Ham.	Gentiaceae	Chirayita	Branches infusion is taken as blood purifier.
33.	*Urtica dioica* Linn.	Urticaceae	Kandali	2-3 drops of root extract are applied in hollow tooth cavity to cure toothache.
34.	*Verbascum thapsus* Linn.	Scrophulariaceae	Ekalveer	The herb is useful for the treatment of asthma.
35.	*Vitex negundo* Linn.	Verbenaceae	Sivali	Leaf infusion is useful in pain with swelling in the ankle or knee joints.
36.	*Woodfordia fruticosa* (L.) Kurtz.	Lythraceae	Dhaula	Flowers are useful in skin diseases and dysentery.
37.	*Zanthoxylum aromatum* DC.	Rutaceae	Timbru	Powder of fruits is used as a remedy for toothache. Small twigs of the branches are used as the stick in toothache.

According to Purohit & Vyas[1], 21000 plant species are useful in the preparation of medicines for various ailments. Herbal medicine is getting more and more popular because it is cheaper and locally available, it brings about a permanent cure for common ailments without side effects. World demand for herbal product has been growing at the rate of 10% to 15% per annum[6]. According to Sandhya *et al.*[7] about 50% of the top prescription drugs in U.S.A. are based on natural products and raw materials.

The indigenous people of Kumaon Himalaya have rich and unique traditional knowledge and practices about the use of natural resources, specially the herbs available in their surroundings for the treatment of various ailments. There is an immediate need for value addition in these sectors in order to save these traditional knowledge from extinction[8]. In recent years, more efforts have been made to document the traditional knowledge of medicinal plants in Uttarakhand State by many workers[2,8-12]. The conservation status of threatened medicinal and folklore plants in Uttarakhand State has also been discussed by Arya & Agarwal[9].

Presently four Govt. and two Non govt. organizations in Almora District are engaged in cultivation and conservation of medicinal plant species like *Acorus calamus, Hedychium spicatum, Ocimum sanctum, Berberis asiatica, Eucalyptus globules, Asparagus racemosus and Swertia chirata etc.* According to Arya & Agarwal[9], *Swertia chirata* is categorized under the Endangered (EN) species, *Cinnamomum tamala* and *Zanthoxylum aromatum are* categorized under the Vulnerable (VU) species. Some species of medicinal plants which are disappearing very fast either because of indiscriminate utilization of their natural habitat in Almora District. So, it is an urgent need to proper documentation and sustainable harvesting for conservation of medicinal plants.

Acknowledgement

This work has been carried out under the UGC-SAP programme, sanctioned to the Department of Zoology and Environmental Sciences, Gurukula Kangri University Haridwar for which we are thankful to University Grants Commission, New Delhi for providing financial support during the period of study.

References

1. Purohit, S.S. and Vyas, S.P. (2005). In : Medicinal Plant and Cultivation., Agrobios India, Jodhpur.
2. Joshi, B.D. (2002). *Him. J. Env. Zool.*, 16(2):233.
3. Kaushik, P. and Dhiman, A.K. (2000). In : Medicinal plants and raw drugs of India. Bishen Singh and Mahender Pal Singh Publication, Dehradun.

4. Sharma, R. (2003). In : Medicinal plants of India - An Encyclopedia, Daya Publishing House, New Delhi.
5. Singh, M.P., Srivastava. J.L., and Pandey, S.N. (2003). In : Indigenous medicinal plants, Social Forestry and Tribals. Daya Publishing House, New Delhi.
6. Sekhar, C., Anjugam, M. and Maheswari, A.M. (2005). *Int. J. For. Usuf. Mngt.*, 6(2):41.
7. Sandhya, B., Thomes, S., Isabel, W. and Shenbagarathai, R. (2006)., *Afr. J. Trad.* CAM, 3(1): 104.
8. Farooquee, N.A. and Nautiyal, A. (1999). *Int. J. Sustain. Dev. World Ecol.*, 6:60.
9. Arya, K.R. and Agarwal, S.C. (2006). Ethnobotany., 18:77.
10. Bisht, N.S., Gera, M., Sultan, Z. and Gusain, M.S. (2005). Indian forester. 131(3):346.
11. Nautiyal, S., Maikhuri, R.K., Rao, K.S. and Saxena, K.G. (2001). Journal of Herbs, Spices and Medicinal Plants. 8(4):47.
12. Singh, Y.P. and Joshi, B.D. (2005). *Him. J. Env. Zool.*, 19(1):97.

22
Haematological Parameters of A Freshwater Teleost, *Heteropneustes Fossilis* (Bloch.)

SATISH PATIL, K. BORANA, S. S. MUDASER ANDRABI AND T. ZAFAR
Department of Applied Aquaculture, Barkatullah University, Bhopal

Abstract

The effects of sub-lethal concentrations of Aldrin were studied on certain haematological parameters of *Heteropneustes fossilis.* For the purpose of this investigation, 10 parameters of *H. fossilis* having mean weight and length 100 g and 18-20 cm, respectively, were used. The RBC count and haemoglobin percentage were found to decrease considerably as compard to normal values. The lymphocytes and eosinophils increased significantly while monocytes and neutrophils decreased.

Introduction

The pesticides are persistent pollutants and are discharged constantly through agricultural, horticultural and municipal effluents into the environment in substantial amount. The survey of literature on pesticide toxicity clearly shows that pesticides cause several haematological, biochemical and reproductive disorders in aquatic organisms[1-3]. The present study was undertaken to evaluate the effect of Aldrin on certain haematological parameters of a freshwater fish, *Heteropneustes fossilis.*

Materials and Methods

Live and healthy specimens of *Heteropneustes fossilis* were collected from the local fish market of Bhopal. They were washed in 0.1% $KMnO_4$ solution and acclimatized for a period of seven days in large aquaria. The length of fish varied from 17 to 19 cm and weight from 18 to 20 gm. Stock solution of aldrin was prepared by dissolving 10 mg in 0.2 ml of acetone and 100 ml distilled water. Further dilutions and desired concentrations were prepared by using standard methods[4].

Acclimatized fish were exposed in three replicates for 15 days at the sublethal concentrations of 0.0004, 0.00026 and 0.0003 ppm of Aldrin. Controls were set separately. Blood parameters were analysed after every sampling and tabulated.

Results and Discussion

The data relating to the effect of Aldrin on haematological parameters are given in Table 1. Significant decrease in RBC and Packed Cell Volume (PCV) was observed in all the concentrations of Aldrin. Significant elevation reduction in haematalogical values of fishes exposed to different environmental toxicants have been reported by several workers. Lai *et al.*[5] reported decreased RBC and HB% in *Heteropneustes fossilis* after 4 days exposure to malathion.

Significant decrease in HB%, PCV, MCV, MCH has been noticed after 15 days treatment with divithion[6]. From the present study, it is clear that Aldrin induces alterations in haematological parameters of *H. fossilis.*

Table 1. Alterations in various haematological parameters of *Heteropneustes fossilis* after 15 days exposure to various concentrations of Aldrin.

Parameters	*Control*	*Aldrin (0.0004)*	*Aldrin (0.00026)*	*Aldrin (0.0002)*
RBC($10^b/mm^3$)	4.67 ± 0.49	4.33 ±0.008	4.41 ±0.006	4.45 ±0.049
Packed Cell Volume (PCV %)	38.50 ± 0.071	35.30 ± 0.022	36.20 ± 0.049	36.50 ±0.006

Values expressed in mean ± S.E.

References

1. Thakur, Nutan and Sahai, S. (1987). In: Changes in blood serum proteins, and other haematological parameters in the teleost, *Channa punctatus* on exposure to organochlorine pesticide, proceedings of the Academy of Environmental Biology, Environ. Pestic. Toxicol. p. 263.

2. Gupta, A.K., Ranjana, A.M. and Dalela, R.C. (1995). *J. Environ. Biol.* 16 (3): 219.
3. Wedemeyer, G.A. and Yasutabe, W.T. (1977). In: Clinical methods for assessment of the effects of environmental stress on fish health. US Fish and Wild Life Services, Washington, D.C. Tech. Paper. p. 89.
4. APHA, AWWA and WPCF (1992). In: Standard methods for the examination of water and wastewater. 18th Ed., American Public Health Association, Washington.
5. Lal, B., Amita Singh, Anita Kumari and Neelima Singh (1986). Biochemical and haematological changes following malathion treatment in the freshwater catfish *Heteropneustes fossilis* (Bl.). Environ. Poll. (Ser A). p. 151.
6. Nath, Rabindra and V. Banerjee (1995). Freshwater Biol. 7 (4): 261.